Cynthia de Faria Pinto
Simone da Silva

Construction Waste Management in the City of Manaus

Cynthia de Faria Pinto
Simone da Silva

Construction Waste Management in the City of Manaus

An analysis from a sustainability perspective

ScienciaScripts

Imprint

Any brand names and product names mentioned in this book are subject to trademark, brand or patent protection and are trademarks or registered trademarks of their respective holders. The use of brand names, product names, common names, trade names, product descriptions etc. even without a particular marking in this work is in no way to be construed to mean that such names may be regarded as unrestricted in respect of trademark and brand protection legislation and could thus be used by anyone.

Cover image: www.ingimage.com

This book is a translation from the original published under ISBN 978-620-6-75614-9.

Publisher:
Sciencia Scripts
is a trademark of
Dodo Books Indian Ocean Ltd. and OmniScriptum S.R.L publishing group

120 High Road, East Finchley, London, N2 9ED, United Kingdom
Str. Armeneasca 28/1, office 1, Chisinau MD-2012, Republic of Moldova, Europe
Printed at: see last page
ISBN: 978-620-6-51783-2

Copyright © Cynthia de Faria Pinto, Simone da Silva
Copyright © 2023 Dodo Books Indian Ocean Ltd. and OmniScriptum S.R.L publishing group

CONTENTS

Construction Waste Management in the City of Manaus - An Analysis from a Sustainability Perspective

Foreword

As a contribution to the city of Manaus, this research aims to demonstrate the current scenario of construction waste management, from the point of view of sustainability, in an attempt to provide subsidies to the public administration for decision-making on the need for actions aimed at minimising impacts on the environment and, consequently, on public health.

Chapter I - A brief introduction

The Industrial Revolution, which began in the 18th century, brought with it industrialisation, technological modernisation, development and economic growth, but also great damage to the environment due to the high consumption of natural resources, as the introduction of new products on the market, produced on a large scale, began to generate a huge variety of solid waste in urban areas (RIBEIRO, 2013).

From the 1970s onwards, concern for the environment began to draw the attention of international authorities to the importance of preserving natural resources and the environment, a fundamental fact for avoiding future problems such as global warming, the extinction of animal and plant species, the greenhouse effect, among others (BRASILEIRO, MATOS, 2015).

The United Nations Conference on Development and the Human Environment, which took place in 1972 in Stockholm, Sweden, was a historic milestone as it was the first major international meeting with representatives from different nations to discuss environmental problems. It led to the drafting of the Stockholm Declaration, with 26 principles, and the creation of the United Nations Environment Programme (UNEP). At the conference, in addition to atmospheric pollution, which was already worrying the scientific community, water and soil pollution from industrialisation and the pressure of population growth on natural resources were discussed (UN, 2020).

This transformation was, therefore, the most relevant and decisive for the promotion of debates, congresses, discussions and, above all, for the creation of laws aimed at raising awareness of the need for more sustainable development, seeking a balance between the conservation and preservation of the environment and technological progress (LOURENÇO, CRISPIM E SILVA, 2015).

In this development process, the civil construction sector has a prominent position in the national economy, being responsible for the largest generation of jobs, both directly and indirectly, but it is the sector that generates the most solid waste and degrades the environment and human health the most, due to its incorrect disposal.

In 2002, the main legal action taken to change this situation was the publication of CONAMA (National Environment Council) Resolution 307. In force since January 2003, this Resolution establishes obligations for generators and municipalities, with the priority objective being the non-generation of waste and, secondarily, reduction, reuse, recycling and final disposal. Another milestone brought about by Law 12.305/2010 was the approval of the National Solid Waste Policy, which mandated municipalities to implement the management of construction waste by drawing up an Integrated Construction Waste Management Plan, thus strengthening the commitment of public and private administration to the continuous improvement of their processes in order to guarantee the environmentally correct disposal of construction waste, to preserve the environment for future generations.

The construction industry is recognised as one of the most important in the world economy, contributing to economic and social development by producing goods whose production chain consumes around 20 to 50% of natural resources, but it is also the industry that pollutes the environment the most through the generation of waste and its inadequate disposal, which accounts for around 72% of the total urban solid waste produced in Brazil (PINTO, 2013).

As well as the unbridled generation of waste, the negative impacts on the environment caused by construction include the emission of polluting gases and high energy consumption, deforestation and the abundant use of water and natural resources (ALLWOOD; CULLEN, 2012).

Construction or demolition waste from the construction industry is known as rubble, which is usually inert material that occupies a large volume in the places where it is disposed of, causing serious socio-environmental impacts, as well as major financial problems for the public administration, which incurs costs for its removal (PINTO, 2013). As such, urgent and effective measures are needed to manage this waste, from segregation at the source to its reuse, recycling and environmentally appropriate and correct disposal, enabling it to be reinserted into the production chain.

In Amazonas, the changes that have taken place since the establishment of the Manaus Free Trade Zone, in terms of population growth and, consequently, its physical structure to meet the demands arising from the growth of the city of Manaus, has increased the generation of construction waste.

It is therefore essential to improve the management of waste generated on public or private construction sites, with a focus on actions aimed at complying with CONAMA Resolution 307/2002 and its updates, as well as applying the guidelines, objectives and targets set out in Law 12.305/2010, which established the National Solid Waste Policy, which seeks to encourage and promote integrated management and management of solid waste, whatever its nature, placing responsibility on public managers to develop projects to promote integrated management and environmentally appropriate management of solid waste in the country.

Chapter II - Historical Approach to Waste

The problem of solid waste disposal began around 10,000 years ago, when man stopped being nomadic and settled down to farm and domesticate animals (VIVEIROS, 2006).

River pollution is an environmental and cultural problem that dates back to ancient times, and the custom of dumping waste in the water continues to this day (JUUTI, 2007).

In the medieval period, the practice of throwing rubbish out of the window into jars was common, meaning that sanitary practices were disregarded. This period was also marked by major epidemics such as smallpox, typhus, cholera, leprosy, among others, since hygiene habits were poorly considered and precarious or non-existent (JUUTI, 2007).

Despite being forbidden, the act of throwing faeces and urine out of windows at night became a habit that was perpetuated in many cities until the 19th century, with repercussions even in Brazil (EIGENHEER, 2009).

> The disgusting task of carrying rubbish and waste from the house to the squares and beaches was usually assigned to the only slave in the family or the one of lower status or value. Every night after ten o'clock, the slaves popularly known as "tigers" would carry tubes or barrels of excrement and rubbish on their heads through the streets (EIGENHEER, 2009).

According to Eigenheer (2009), innovations in urban cleaning began in the 14th century with the use of carts. A regular rubbish collection and street cleaning service was set up in Europe, under the responsibility of private individuals. In Brazil, on 11 October 1876, the firm of Aleixo Gary was contracted to carry out public cleaning and was considered an important milestone for urban cleaning in Rio de Janeiro. Hence the name "gari" for some urban cleaners to this day.

It wasn't until the second half of the 19th century that a clear distinction was made between rubbish (solid waste) and sewage (faeces, urine, etc.), when the latter began to be collected separately through sewage systems; great importance was attached to water quality and the need to separate sewage from solid waste was established (Eigenheer, 2009).

In the 18th century, industrialisation and technological modernisation took place, but public health and basic sanitation also became important factors in the production process (REZENDE AND HELLER, 2008).

From the second half of the 19th century onwards, the first concepts of waste treatment emerged, with incinerators and models of sorting plants in Europe, and in the United States the system of selective collection began. However, despite the innovations in waste collection and improvements in waste reuse, final disposal is still a major problem, as waste is still disposed of in rivers, seas and neighbouring areas (EIGENHEER, 2009).

Chapter III - Sanitation and the National Solid Waste Policy

Since the 16th century, Brazil has dealt with waste management in a fragmented way, incorporating this service and public cleaning into basic sanitation, which has somewhat hindered the less favoured population's access to essential collective health and well-being services. In Brazil, the historical evolution of sanitation was influenced by the customs of ancient peoples, who only began to worry about waste generation, public health and sanitation when epidemics broke out. It was believed that diseases were punishments from the deities and that actions should be taken to treat waste (REZENDE E HELLER, 2008).

From the 16th to the 19th century, sanitation actions were restricted to the wealthiest cities where the elite were concentrated. The history of solid waste in Brazil has always been geared towards economic interests (REZENDE E HELLER, 2008).

From 1970 onwards, the National Sanitation Plan (PLANASA) was implemented, but by 1990 it had declined (REZENDE E HELLER, 2008). In 1981, the National Environmental Policy (PNMA) was instituted through Law No. 6.938 of 31 August 1981. Although it is considered one of the greatest advances in Brazilian environmental legislation, waste only began to be treated in a special way after the 1988 Federal Constitution was enacted (SILVA et al., 2017). From then on, municipalities became the owners of urban cleaning services and, consequently, solid waste management became their responsibility.

In 1998, Law No. 9.605, known as the Environmental Crimes Law, was sanctioned, which provides for criminal and administrative sanctions arising from conduct and activities harmful to the environment. Articles 54 and 56 state that it is a criminal offence to cause pollution through the improper disposal of waste, and that anyone who handles hazardous waste in breach of the provisions of the law or regulations is an offender.

In 2003, the National Secretariat for Environmental Sanitation (SNSA) was created as the body responsible for monitoring and disseminating information on the collection and treatment of solid waste in various Brazilian cities, through the National Sanitation Information System (SNIS) (BRASIL, 2008), whose order of priority is: non-generation, reduction, reuse, recycling, treatment of solid waste and environmentally appropriate final disposal of waste (BRASIL, 2010).

However, in the 1990s, the national guidelines for Basic Sanitation were approved, with Law No. 11.145 of 2007, considered one of the milestones that emphasise the great importance of proper waste management, with a view to urban cleaning and solid waste management, which are aspects of sanitation (BRASIL, 2007).

Throughout history, sanitation and public health have been interlinked, but often on opposite sides. Water supply was prioritised, but sewage, public cleaning and rainwater drainage were considered less important (MOTA E SILVA, 2014).

The regulation of solid waste occurred very slowly in Brazil and, even though it began in the 1980s, it was only 20 years later that the National Solid Waste Policy (PNRS) was approved through Law 12.305/2010, a fact that highlights the lack of prioritisation on the part of the public authorities (BRASIL, 2010). It should be noted that the regulation of this law was published on 12 January 2022, through Decree 10.936/2022. Given the need to implement public policies aimed at guaranteeing the proper disposal of solid waste, the introduction of Law 12.305/2010 has become a historic milestone for Brazilian environmental policy, of paramount importance for solid waste management, given the potential damage caused to the environment.

To this day, waste management is a major problem for public authorities, as some Brazilian municipalities dispose of waste in inappropriate places, causing serious risks to the environment and public health. The disposal of waste in rubbish dumps is still common practice in some places, which violates the principle of the PNRS, which requires that these dumps be eliminated or replaced by alternatives that have less impact on the environment, such as sanitary landfills, and that the waste be correctly disposed of (Law 12.305/2010). The law required the closure of rubbish dumps in municipalities by 2014, under penalty of a fine under Law 9605/98 on environmental offences, which did not happen. A new deadline was set for the end of rubbish dumps in Brazilian municipalities, through the Sanitation Framework, sanctioned in July 2020 (Law No. 14.026, of 15 July 2020), which varies according to the existence of solid waste plans and the number of inhabitants in the cities and which, in general, provides for the closure of all rubbish dumps in Brazil by 2024 (AGÊNCIA BRASIL.EBC.COM.BR).

The National Solid Waste Policy (PNRS) is governed by Law 12.305/2010 in Brazil and provides for principles, objectives and instruments, as well as all types of waste (industrial, domestic, healthcare, electronics, etc.), and also determines the guidelines for integrated management - uniting all those involved, proposing a sharing of responsibility for the life cycle of products, involving consumers, manufacturers, distributors and others. It also establishes co-operation between federal, state and municipal authorities, society and the industrial production sector, with the aim of seeking alternatives to the country's environmental problems and managing solid waste in the best possible way.The main objective of the PNRS is reduction, i.e. the non-generation of waste, through its treatment and reuse. This will increase recycling in the country and reduce the use of natural resources such as water and energy in the production of new products. In other words, the PNRS encourages the promotion of actions aimed at the non-generation, reduction, reuse, recycling, environmentally correct treatment and final disposal of waste, industrial recycling, the use of clean technology, integrated solid waste management and ongoing training (BRASIL, 2010). In addition to the issues mentioned above, the PNRS also addresses other issues such as reverse logistics, selective collection, integrated solid waste management plans and tax incentives (ELK and BOSCOV, 2020).

The PNRS defines solid waste as "any discarded material, substance, object or good resulting from human activities in society". Disposing of this waste does not mean that it no longer has value, but rather that it is no longer needed by the person who disposed of it. However, there is a good chance that this waste will still be useful to other people, either in its original form or transformed (PORTO, JUNIA DE OLIVEIRA, 2017).

Articles 16 and 18 of Law 12.305, of 2 August 2010, state that the preparation of state and municipal solid waste plans, under the terms set out in this law, is a condition for states, municipalities and the Federal District to have access to federal funds, or funds controlled by the federal government, for projects and services related to solid waste management, or to benefit from incentives or funding from federal credit or development entities for this purpose.

The National Solid Waste Management Information System (Sistema Nacional de Informações sobre a Gestão dos Resíduos Sólidos - SINIR) is one of the instruments of the National Solid Waste Policy (Política Nacional de Resíduos Sólidos - PNRS), established by Law No. 12.305, which collects, integrates, systematises and makes available data on the operation and implementation of solid waste management plans.

The sole paragraph of Law 12.305, of 2 August 2010, **requires** states, the Federal District and municipalities to provide the federal body responsible for coordinating SINIR with all the necessary information on waste within their sphere of competence, in the manner and at the intervals established by regulation.

According to the PNRS, there is a distinction between waste and rejects. The latter cannot be economically treated or recovered. They must therefore be disposed of in an environmentally appropriate manner.

Solid waste is classified according to its origin and hazardousness (BRASIL, 2010, Art. 13). And art. 13, I of Law 12.305 of 2010 provides for the classification of solid waste according to its origin and establishes:

a) Household waste: waste originating from domestic activities in urban residences;

b) urban cleaning waste: waste originating from sweeping, cleaning public places and streets and other urban cleaning services;

c) solid urban waste: the waste listed in points "a" and "b";

d) waste from commercial establishments and service providers: waste generated in these activities, with the exception of that referred to in points "b", "e", "g", "h" and "j";

e) waste from public basic sanitation services: waste generated by these activities, with the exception of that referred to in point "c";

f) industrial waste: waste generated in production processes and industrial facilities;

g) health service waste: waste generated in health services, as defined in regulations or standards established by SISNAMA and SNVS bodies;

h) civil construction waste: waste generated in the construction, renovation, repair and demolition of civil construction works, including waste resulting from the preparation and excavation of land for civil works;

i) agroforestry waste: waste generated in agricultural and forestry activities, including waste related to inputs used in these activities;

j) waste from transport services: waste originating in ports, airports, customs, road and rail terminals and border crossings;

k) mining waste: waste generated during the research, extraction or processing of minerals.

As for dangerousness, Law No. 12,305 of 2010 states:

a) hazardous: those which, due to their characteristics of flammability, corrosivity, reactivity, toxicity, pathogenicity, carcinogenicity, teratogenicity and mutagenicity, present a significant risk to public health or environmental quality, in accordance with the law, regulation or technical standard;

b) non-hazardous: those not covered by point "a".

Chapter IV - Construction Waste - Legal and Regulatory Aspects

Concern about environmental issues became relevant, evident and necessary after the Industrial Revolution, when there was a significant increase in the generation of solid waste due to growing urbanisation, population growth and, above all, the growth in production to meet new demands. All these factors led to serious public health problems (MOTA, 2014, p.16).

Every year, around 3 billion tonnes of waste are thrown away, of which approximately 90 million tonnes are hazardous waste. This figure is equivalent to 6 tonnes of solid waste for every person per year (FARIAS, 2013).

This means that new dynamics need to be incorporated in order to achieve effective environmental protection, stop using inappropriate practices and enable the transition towards a circular economic model and guarantee the universalisation of proper solid waste management. For this to happen, it is essential to know the current waste management system, the characteristics of discarded materials and the main indicators of service provision (generation, collection and disposal of waste, available resources, existing initiatives, among others), in order to guide best practices and make solutions feasible, according to demand.

The Brazilian Association of Technical Standards (ABNT), through Brazilian Standard (NBR) 10.004, in item 3.1, defines solid waste as solid and semi-solid waste resulting from industrial, domestic, hospital, commercial, agricultural, service and sweeping activities. This definition includes sludge from water treatment systems, sludge generated in pollution control equipment and installations, as well as certain liquids whose particularities make it unfeasible to discharge them into the public sewage system or bodies of water, or require solutions that are technically and economically unfeasible in view of the best available technology. This classification shows the diverse amount of waste present in society and how important it is to improve collection, treatment and final disposal processes.

The Brazilian Association of Technical Standards (ABNT) classifies solid waste according to its potential risks to the environment and public health, so that it can be managed properly. They are classified as follows:

a) Class I waste - Hazardous;

b) Class II waste - Non-hazardous;

b1) Class II A waste - Non-inert;

b2) Class II B waste - Inert.

Concern about the generation of waste has been increasing considerably over the last few decades on the national and international stage, given the widespread dissemination of the consequences brought about by economic development, broadening environmental awareness to a certain extent and alerting the world to the need for urgent action to minimise the negative impacts that the planet is suffering (RIBEIRO,2013).

With the creation of the National Environment Council (CONAMA), through Federal Law No. 6.938/81[1], a Brazilian collegiate body responsible for adopting measures of an advisory and deliberative nature regarding the National Environment System, it was possible to put into practice actions relating to the generation of construction waste, through Resolution 307 of 5 July 2002. Over the years, CONAMA Resolution 307/2002 has been amended by Resolutions 348/2004, 431/2011, 448/2012 and 469/2015.

CONAMA Resolution 307/2002 has become a milestone in the environmental context. In its Article 5, CONAMA establishes the Municipal Construction Waste Management Plan as an instrument for implementing construction waste management, to be drawn up by municipalities and the Federal District, in line with the Municipal Integrated Solid Waste Management

Plan. Article 8 states that Construction Waste Management Plans must be drawn up and implemented by large generators and will aim to establish the necessary procedures for the environmentally appropriate handling and disposal of waste.

According to Article 3 of the Resolution, construction waste is classified as described in Table 1.

Table 1 - Classification of Construction Waste.

CLASS	DESCRIPTION
A	Waste that can be reused or recycled as aggregates, such as:
	a) construction, demolition, renovation and repair of paving and other infrastructure works, including earthworks;
	b) construction, demolition, renovation and repair of buildings: ceramic components (bricks, blocks, tiles, cladding panels, etc.), mortar and concrete;
	c) the process of manufacturing and/or demolishing pre-moulded concrete parts (blocks, pipes, kerbs, etc.) produced on construction sites;
B	This is waste that can be recycled for other uses, such as plastics, paper, cardboard, metals, glass, wood and others;
C	This is waste for which no economically viable technologies or applications have been developed that would allow it to be recycled/recovered, such as gypsum products;
D	This is hazardous waste from the construction process, such as paints, solvents, oils and others, or contaminated waste from demolition, renovation and repair of radiological clinics, industrial facilities and others.

Source: CONAMA Resolution 307/2002.

According to Resolution 307/2002-CONAMA, construction or demolition waste from the construction industry is known as rubble. They are usually inert materials that occupy a large volume in the places where they are disposed of, causing serious socio-environmental impacts, which is why urgent and effective measures are needed to manage this waste, from segregation at the source to its reuse, recycling and final, environmentally appropriate and correct disposal, with the disposal of such waste in MSW landfills, "dumping grounds", hillsides, bodies of water, vacant lots and in areas protected by law being prohibited (BRASIL, 2002).

Figure 1 shows the Consolidation of Legislation on the classification of construction waste according to its origin and potential risks to the environment and public health.

Figure 1 - Consolidation of legislation on the classification of construction waste.

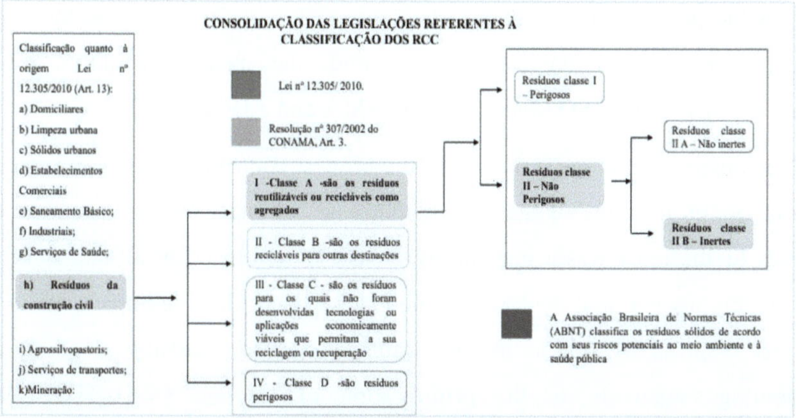

Source: Own authorship.

Chapter V - Environmental Impacts of Construction Waste

Environmental impact, according to Conama Resolution N°001 of January 1986, is defined as any alteration to the physical, chemical and biological properties of the environment, caused by any form of matter or energy resulting from human activities that directly or indirectly affect the health, safety and well-being of the population; social and economic activities; biota; the aesthetic and sanitary conditions of the environment and; the quality of environmental resources (OLIVEIRA, SOARES, QUARESMA and ADORNO, 2020).

The construction industry is one of the biggest contributors to environmental degradation, as it is one of the sectors that uses the most natural resources and disposes of the most waste improperly. Waste is generated at all stages of the construction process, and the health issue receives considerable impact due to the incorrect disposal of rubble in the drainage network, causing flooding and contributing negatively to the proliferation of diseases (VALOTO, 2007; OLIVEIRA, SOARES, QUARESMA and ADORNO, 2020).

Among the main impacts on the environment and public health caused by incorrect and inadequate disposal of construction waste are changes to the landscape, degradation of water sources, obstruction of drainage systems, proliferation of disease-causing vectors (malaria, yellow fever, dengue fever), contamination of the soil with harmful materials (paints, solvents, fibre cement, etc.), silting up of watercourses and difficulty in moving vehicles on public roads and people on pavements (Figure 3 A and B).), silting up of watercourses and difficulties for vehicles on public roads and people on pavements (Figure 3 A and B) (MOTA; SILVA, 2014).

Figure 1 - Environmental Impacts of Irregular Disposal.

Source: Blog Fim do Lixo (2022) Source: SEMULSP 2021 - Igarapé do 40 (Manaus/AM)

Technological development and the advancement of science have brought post-modern man conflicts arising from his own creativity and technological evolution, due to the environmental consequences resulting from this innovation and modernisation, such as the problem of difficult-to-decompose waste and its toxicity (Costa, 2013).

The basic composition of rubble from construction and infrastructure work can vary according to the characteristics of the project, the materials available in the region, the technologies applied and the quality of the labour used in the building systems (Wada, 2013).

The waste generated contains various materials such as: asphalt, glass, concrete, mortar, lime, ceramic material, pruning material, crushed stone, wood, blocks and bricks, paper, paints and varnishes, plaster, plastics, solvents, pigments and soil and the risks can be present in various forms, especially in chemical substances, physical, mechanical and biological agents (Nascimento, 2013; Porto, 2015).

Law 12.305/2010 defines the most appropriate actions and destinations for waste, as well as for tailings. According to the PNRS, waste is solid waste which, after all possibilities of treatment and recovery have been exhausted in accordance with available and economically viable technologies, has no other possibility than environmentally appropriate final disposal. This is the last alternative to be adopted by the generator.

An important precaution that must be taken into account is that technological evolution occurs very quickly, so what may be considered waste today may be waste tomorrow. This process should be considered changeable and not static (PORTO, JUNIA DE OLIVEIRA, 2017).

Distributing and sorting waste in sanitary landfills, in compliance with specific operational rules, in order to avoid damage to public health and safety, minimising environmental impacts, is the environmentally appropriate way to dispose of it (BRASIL, 2010a).

The purpose of some waste treatment processes is to utilise waste and others to treat waste, while final disposal only involves disposing of the waste (SILVA FILHO E SOLER, 2013).

It is considered inadequate disposal when it takes place in open dumps or controlled landfills. This is because, although there is a vegetation cover, when rubbish is deposited directly on the ground, without undergoing any operation and control, it contaminates the soil and groundwater through the leachate produced, causing serious harmful effects to public health and the environment (BRASIL 2010).

Despite the PNRS prohibiting the open disposal of waste in seas, lakes and rivers, causing environmental contamination, damaging the quality of air, water and soil and public health, and imposing fines and penalties on offenders, the improper disposal of waste is still prevalent in Brazil (POZZETTI and CALDAS, 2019).

There is a need to rethink waste disposal and invest in new production process technologies, with the aim of continuously improving current processes, using fewer natural resources and reusing waste as much as possible.

Table 2 shows the types of final waste destination and their characteristics:

Table 2 - Types and Final Destinations of Solid Waste and their Characteristics.

Type	Features
Open dumps or rubbish dumps	It receives untreated rubbish, either in vacant lots or through the collection system, without any care for the environment or public health, encouraging marginalised scavenging activities.
Dumps in flooded areas	Dumping rubbish in mangroves, swamps, lakes, rivers, streams, seas, etc.
Controlled landfill sites	Destined to receive rubbish, it has a daily cover, but generates liquid and gaseous by-products such as leachate and methane.
Landfill sites	A suitable site for the final disposal of rubbish, with waterproofing, drains for leachate, gases and rainwater, daily covering and compaction of rubbish.
Transhipment centres	Places where collection trucks dump waste to optimise collection through pressing and to minimise costs cannot be considered final disposal.
Sorting Centres	Places for sorting rubbish (metals, glass, paper and plastics), which can be done by conveyor belt or other equipment, for commercial purposes and the organic matter generated can be composted.
Composting centres	Intended for aerobic decomposition to produce soil conditioner.
Incineration centres	Places where the controlled combustion of carbon waste (food waste, garden rubbish, plastics and paper) takes place, allows gaseous effluents, ash and slag to be reused. It has been an alternative for healthcare waste, although it is being replaced by cheaper and more efficient techniques.

Source: adapted from (MANCINI et al., 2012)

Chapter VI - Waste and the Generation of Construction Waste

The high level of waste is one of the main characteristics resulting from the poor management of activities inherent in the construction sector, as well as the great impact of the volume of waste generated and raw materials consumed (PIMENTEL, 2013).

In a construction project, waste represents an average of 5% and this percentage does not include the informal market, which is responsible for more than half of all constructions. While this waste is not so great in percentage terms, financially the figure is different, since the cost of the work always involves large sums (FILHO, 2015).

As well as damaging the environment, construction waste causes logistical problems and financial losses (NAGALLI, 2014; SILVA, SANTOS and ARAÚJO, 2017; HOLANDA et al., 2018).

Construction waste is heterogeneous, made up of various materials, whether they result from the demolition of buildings or infrastructures, or from losses during construction, i.e. leftover and wasted materials from construction stage activities (ALMEIDA et al., 2015).

Waste and loss can at first be confused, as they seem to represent the same product, however, one derives from the other, and loss only materialises as waste when it is not absorbed during construction (BRAGA and VEIGA, 2017).

From the project conception phase to the execution of the work, all stages must be managed (ALMEIDA et al., 2015).

The main causes of waste generation (Figure 4) can be related to flaws in the design, management and planning of the work (OSMANI, 2011).

Figure 2 - Sources of Waste in Construction.

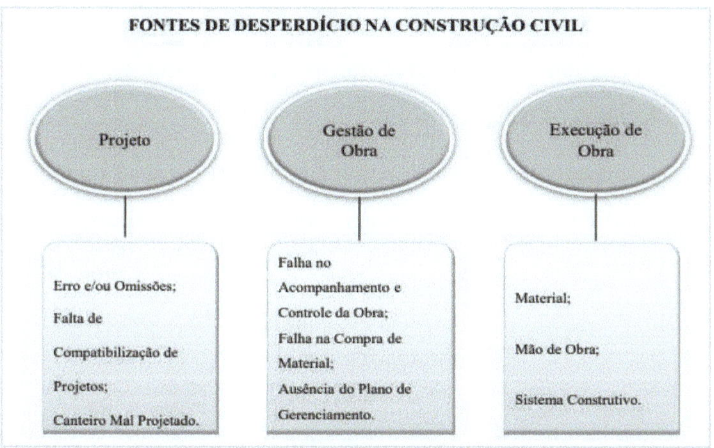

Source: Own authorship

Project Development Phase

The major cause of waste on the building site comes from the design phase. The lack and/or poor quality of projects, technical specifications, incompatibility, lack of detailing, failure to plan based on projects and the schedule for purchasing materials and changes to projects are largely responsible for waste on site (BRAGA and VEIGA, 2017). Design errors, faulty memoranda, lack of detailed design and various other factors end up further increasing the generation of this waste (OLIVEIRA, 2013).

It is important to study, design and implement the construction site with the aim of fully developing the work, avoiding improvisation, displacements throughout the work, unnecessary crossings in the flow of operations and conflicts, thus avoiding the waste of material and labour (NOBRE, 2015).

Braga and Veiga (2017) propose the following design practices for construction sites to reduce waste:

- Designing the site layout to avoid unnecessary transport;

- The construction site project must have a systemic approach, covering the entire production cycle: internal transport, proper storage of materials and waste, location of equipment, etc;

- Specify signposts in the construction site project because signposted works reduce waiting times for materials, unnecessary transport, accidents and wasted materials.

Construction management

Failure to monitor and control the work: This occurs when a service is carried out with technical specifications that fall short of the project requirements, which leads to rework or defective products. This type of loss usually occurs due to a lack of harmony between design and execution, a lack of planning or a lack of control during the execution phase. When there is inefficiency in administrative management, which emphasises correcting problems rather than preventing them. This is due to the managers' lack of involvement in the production process. Examples include ceramic tiles peeling off, painting and waterproofing failures, oversizing structural parts or using concrete with a higher fck than necessary or coating with a greater thickness than necessary (BRAGA E VEIGA, 2017).

Failure to Purchase Materials and Equipment: The concept of waste management is most often related to the waste of materials in construction, however, the meaning extends to any inefficiency that is reflected in the use of materials and equipment, in quantities greater than those required for production (SANTOS, 2016). Therefore, losses involve both the disposal of waste and the wastage of materials in the execution of unnecessary services that generate costs and do not add value. This is due to the lack of links between construction work and support activities, such as purchasing, stocks and maintenance. Planning the purchase of materials, avoiding oversized purchases and overstocking, and planning the schedule in as much detail as possible, taking into account all aspects of the work, construction process, materials used, internal transport, inspections, among others, must be considered and analysed in order to avoid waste during the execution of the work (BRAGA and VEIGA, 2017).

Absence of a Management Plan: It is of the utmost importance to observe and follow the Waste Management Plan, considering the hierarchical stages of reduce, reuse and recycle. Other important actions are needed to reduce material waste and financial losses due to these losses (BRITO, 2015).

The National Solid Waste Policy Law stipulates that all construction companies must draw up a Waste Management Plan. This plan will comply with the provisions of the Municipal Integrated Solid Waste Management Plan of the municipality in which the work is being carried out.

Solid waste management, according to Law 12.305/2010, is the set of actions carried out, directly or indirectly, in the stages of collection, transport, transhipment, treatment and appropriate final disposal of solid waste and environmentally appropriate final disposal of waste.

The Solid Waste Management Plan is a document that identifies the type and quantity of waste generated, as well as the environmentally correct practices for handling, packaging, transport, transshipment, treatment, recycling, destination and final disposal, with the consequences of minimising environmental impacts if applied correctly. The absence of a plan, non-observance of procedures or failure to comply with them can be considered one of the causes of waste, in addition to that which occurs during the operational phase of the work (MMA, 2015).

The generation of construction waste (CCW) is mostly due to the waste of building materials at the time of construction. But there are also losses in the processes of receiving, transporting and storing materials (LIMA and LIMA, 2012).

Low-skilled labour and various other factors end up further increasing the generation of this waste (OLIVEIRA, 2013).

Material

This group includes basic materials such as sand, cement, pebbles and others. In the group of materials that should have a cutting plan so that there is no waste are pipework, steel, wood, ceramics, among others (PALIARI, 1999).

Another very important measure within this process that should be adopted is the replacement of conventional materials with new ones that are technologically better and generate less waste (MICHEL, 2010).

Waste in the construction industry is a reality and unnecessary losses can be minimised or even avoided with actions aimed at saving inputs and reducing waste. Direct loss occurs when wasted material can no longer be recovered or used for its intended purpose, i.e. that applied during the execution of a service (MATUTI and SANTANA, 2019).

Concrete waste is present in infrastructure and construction work and, among the other waste products, it is the most representative in the chain. Due to its application at various stages, whether in dosing, production, application or technological control, concrete produces a considerable amount of waste. It is therefore an input that generates waste and financial loss (BULHÕES, 2014).

Possible forms of material waste in the construction industry are identified during the execution of the work and can occur through overproduction, longer waiting times, transport, losses in the process itself, improper storage of material and losses in movement (KARPINSK et al., 2009).

Overproduction: These losses are due to the production of quantities in excess of what is required for the process, at a stage before its application, i.e. during preparation, due to errors in dosage and overproduction. Examples include mortar or concrete made in excess of the quantity required for the job.

Longer waiting times: These are related to the lack of synchronisation between the production of employees and the receipt of materials or equipment. This loss leads to production stoppages due to a lack of material or equipment to carry out a particular activity.

Transport: Within the construction site, there are various forms of material waste, such as handling and transport, which can occur due to poor distribution on the site, difficult access, excessive handling or inadequate transport equipment.

When a construction site has an inefficient organisation of its physical space or poorly programmed activities, there are transport losses, which occur when a material is handled more than necessary. An example of this is the excessive time taken to move a material from the stockroom to the lift, when they are too far apart, or the breaking of bricks or blocks along the way" (NUNES and SOUZA, 2018, p. 46).

Processing losses: These are caused by the very nature of carrying out an activity, poor execution, lack of standardised procedures, inefficient working methods, lack of training for the workforce or deficiencies in project detailing. Examples include the breaking of walls and slabs in order to install electrical and plumbing installations, and the manual breaking of blocks due to a lack of half blocks (KARPINSK et al., 2009).

Improper storage: When bulk materials are stored improperly, they can be washed away by rain or de-characterised by the weather (CAMENAR and SCHEID, 2016). They are also associated with the existence of excessive stocks due to the inadequate scheduling of material deliveries, which can lead to situations where there are no suitable locations, causing losses due to cement deterioration as a result of storage in contact with the ground and/or in piles that are too high (KARPINSK et al., 2009).

Movement losses: These losses are due to the poor planning of the physical layout of a construction site, where work fronts are far apart or difficult to access, generating unnecessary movement of workers during the execution of their activities (KARPINSK et al., 2009).

Labour

Lack of knowledge and awareness among the workforce, especially about the reduction or non-generation, reuse, sorting and characterisation of waste, are also reasons for the waste generated on site. It is very important, in addition to investment in construction planning, to invest in training to qualify the workforce, thus facilitating the execution of tasks and increasing the level of knowledge and awareness, so that the service is carried out with quality, in less time and with fewer losses (BRAGA and VEIGA, 2017).

Lack of Qualification: Compared to other industrial sectors, the construction industry is lagging behind due to low productivity, lack of quality control, delays and a high rate of material waste (CHAGAS, PADILHA JR and TEIXEIRA, 2015).

Rework: Criteria must also be observed for the selection of qualified labour to carry out tasks in the various stages of the project, thus avoiding rework and increased costs (ALMEIDA et al., 2015). Examples include waterproofing, poorly executed painting, laying ceramic tiles and plastering, as processes subject to losses at the application stage (SOUZA, 2014, p.59).

Lack of Standardisation: Waste during the execution of the service is also caused by the manufacture or use of non-standard products, unskilled labour and, as a consequence, rework and a drop in final performance. Services such as painting, laying ceramic tiles and plastering should be checked before they are applied (SOUZA, 2014, p. 59).

Services should be simplified by standardising procedures and organising tasks to avoid unnecessary activities (BRAGA and VEIGA, 2017).

Conventional Building System

Reducing waste generation is directly related to the construction system used, implying environmental and economic benefits. In Brazil, the conventional construction system is predominantly used (BRAGA; VEIGA, 2017).

The conventional structural system (figure 5) consists of reinforced concrete beams, slabs and columns, which are made up of a combination of fresh concrete and steel bars. These structural components are moulded and braced until they reach the necessary strength to support themselves, forming the structure of the building (NUNES; SOUZA, 2017).

Figure 3 - Conventional Building System.

Source: Jornal de Brasília - 2021

Reinforced concrete is widely used in the construction industry and has a significant value within the construction site due to its high utilisation. It is used to build houses, low-rise and high-rise buildings, bridges, viaducts, tunnels, sheds, silos, towers, roads and many other projects. Its applications range from foundations, pillars, beams and slabs to blocks, roof tiles and tanks, among countless other applications (NUNES, SOUZA, 2017). Foundations in the conventional system depend on the type of soil, the construction and its load. The structural elements have a high self-weight, so their foundations need to be more robust (MIRANDA et al., 2022).

The structural phase of a building is the one that involves the largest financial percentage, ranging from 29.2 to 35.7 per cent of the total cost of the work and, theoretically, the greater the weight that the structure has to support, the greater the structural loading and, consequently, the greater the amount spent on this phase (MATTOS, 2006).

In this system, ceramic blocks are used for the closure, and the wall usually needs to be covered with a layer of plaster, render and rendering to protect the structure and its interior from the weather. This process is slow and consumes a large volume of materials (GRUBLER, 2021).

The main function of conventional masonry is to adapt and establish separation between rooms and it must be resistant to humidity and thermal movements; resistant to wind pressure and resistant to rainwater infiltration; thermal and acoustic insulation; control of water vapour migration and regulation of condensation; a base or substrate for coatings in general and safety for users and occupants (BARROS, 2009).

Plumbing and sanitary installations are similar to electrical installations, which means that when these installations pass through the walls, cuts have to be made, generating a large volume of rubble (CRUZ, 2021).

In roof structures, wood is the material most often used to make the trusses, angles, struts and other components that support the roof tiles, with ceramic and fibre cement tiles being the most commonly used (CASSAR, 2018). The characteristics of this construction method (Table 3) are its low productivity and high waste rate (CRUZ, 2021).

Table 3 - Physical and Visual Characteristics of Vertical Sealing.

Vertical sealing	Conventional Masonry
Materials	Baked clay brick with holes, lime mortar, sand and cement.
Execution	Preparing mortar, laying one brick on top of another, maintaining plumb, plastering and rendering. Time-consuming, with a high level of waste.
Finishing	It requires a lot of care, as the plaster contains some irregularities and is rough.
Labour	A larger number of workers because it covers several stages.
Weight	Thickness of 15 cm = 225 to 270 Kg/m2.

Use	It can be used in any environment.
Time	15 to 20 m^2 , per day.

Source: adapted from Anjos & Teixeira (2017)

In some regions, this conventional construction method is completely handmade, and its characteristics are low productivity and a high level of material waste, due to the fact that it is produced *on site*, thus slowing down the execution of the project. As for the labour force, much of it is generally unprepared, which leads to excessive waste and rework (PRUDÊNCIO, 2013).

The speed of the conventional construction system is compromised due to the low level of industrialisation, as well as the high demand for time and waiting, due to the materials used, such as concrete and mortar, needing time to dry and cure. Another important factor is the dependence on the completion of one stage for the start of another, and rework is also part of this system (ALVES, 2015).

Table 4 shows the advantages and disadvantages of the conventional masonry construction system.

Table 4 - Advantages and Disadvantages of the Conventional Building System.

ADVANTAGES	DISADVANTAGES
It can be used on construction sites with large spans.	The plumbing and electrical installations are done after the masonry, causing waste because the walls are broken down and then sealed with mortar, resulting in rework.
It makes it possible to build large projects because its "skeleton" or structure is made of reinforced concrete.	Lack of quality materials and poor execution.
Good fire resistance.	High rate of water wastage.
Durability of more than a hundred years without protection and maintenance.	The execution time is usually longer, causing delays in the delivery of the work.
Fewer architectural design constraints.	A lot of dirt and rubble accumulates on the site.

| Easy to find materials for its execution. | Lack of skilled labour. |

Source: Salomão, Soares, Lorentz, Larissa (2019)

Studies carried out by USP's Polytechnic School concluded that material losses amount to 8% and financial losses, including those related to rework costs, amount to 30%. To combat these problems caused by waste, it is necessary to seek constant improvements in the quality of the work, investing in technologies in order to achieve cleaner constructions (CONSTRUÇÃO, 2018).

The two stages that generate the most CDW are the superstructure and cladding, which together account for 81 per cent of the waste generated during construction (Figure 6) (MARQUES, OLIVEIRA and PICANÇO, 2013).

Figure 4 - Construction Waste, by Construction Stage.

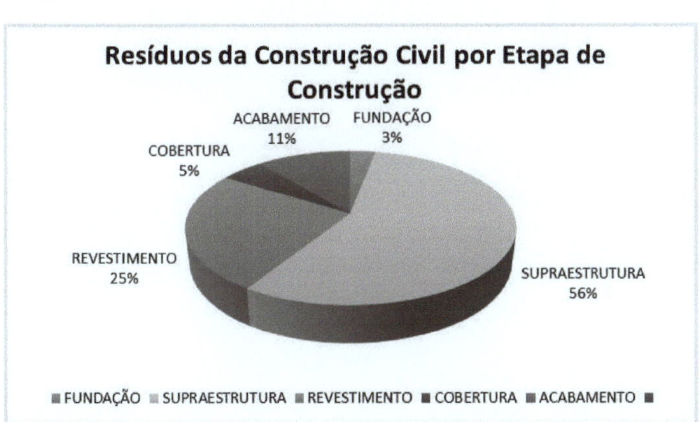

Source: (Marques, Oliveira and Picanço, 2013)

As for the composition of the waste, Marques, Oliveira and Picanço (2013) concluded in their study that most of the waste generated corresponds to mortar and ceramics (Figure 7).

Figure 5 - Composition of Construction Waste.

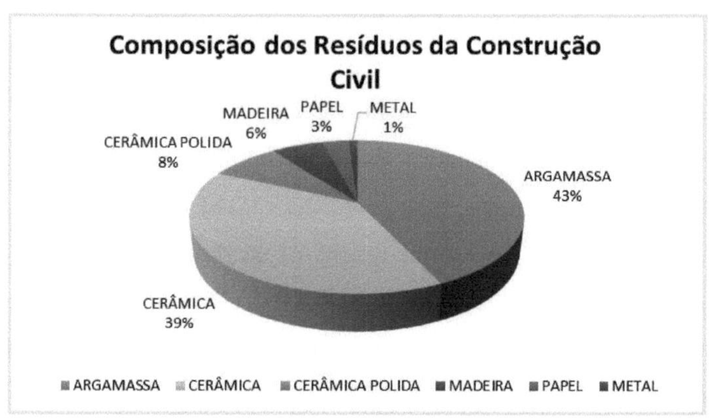

Source: (Marques, Oliveira and Picanço, 2013)

Chapter VII - Construction Waste and Sustainability

One way for society to tackle the problem of waste generation, from production to disposal, would be to adopt the policy of reduction, reuse and recycling, known as the 3R's (ANDRADE, 2014). The policy was first used in 1992 at the Earth Summit in Rio de Janeiro, with the aim of adopting actions for sustainable development. The main objective of the policy is to sensitise the population to become properly aware of urban and industrial waste management (ALKMIM, 2015).

> "Sustainable development is development that meets current needs without compromising the ability of future generations to meet their own needs (UN, 1983)."

The PNRS establishes the following guidelines in its Article 9: In the management and management of solid waste, the following hierarchical order of priority must be observed, as shown in Figure 8: non-generation, reduction, reuse, recycling, treatment of solid waste and environmentally appropriate final disposal of waste.

Figure 6 - Hierarchy in CCW management.

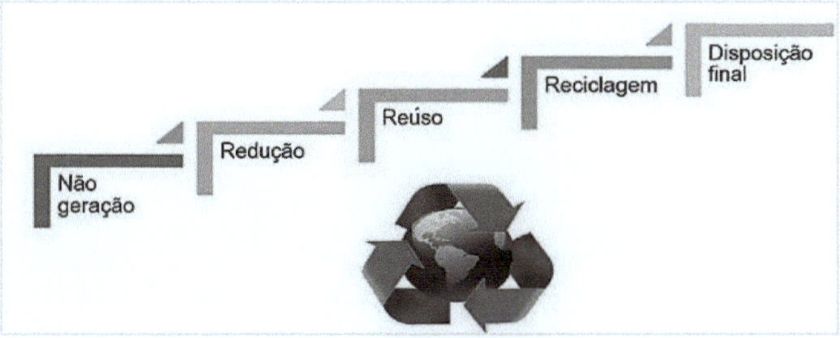

Source: Waste hierarchy, according to Hewerton Bartoli - ILLUSTRATION "Planning and Management of Construction and Demolition Waste" study

Figure 9 shows possible solutions to be adopted in the construction industry, with a view to reducing waste, mitigating waste generation, increasing productivity and the quality of the work and, consequently, minimising negative impacts on the environment through sustainable tools applied to the construction industry.

Figure 7 - Reducing Waste Generation with a Focus on Sustainability.

Source: Own authorship

Chapter VIII - Reverse Logistics in Construction

Reverse logistics can be defined as the area of business logistics that plans, operates and controls the flow and corresponding logistics information of the return of after-sales and consumer goods to the production cycle, through reverse distribution channels, adding value of various kinds: economic, ecological and social (CARNEIRO; MARTINS, 2019).

The PNRS considers it essential that reverse logistics be included in the industrial process. In this way, a set of actions will be established between those involved in the life cycle of a product (from industry to shops), with the aim of returning waste to its generators for reuse in new products or to be correctly disposed of (BRASIL, 2010a).

In reverse logistics, responsibility for the proper treatment of waste and refuse is shared between society, the public authorities and the private sector. From this point of view, the collection and return of waste is the responsibility of the product manufacturer, whose main aim is to ensure that disposal is as environmentally appropriate as possible (MMA, 2015).

Data released by the Brazilian Reverse Logistics Council (CRLV) and the Brazilian Logistics Association (Aslog) showed that reverse logistics currently generates around 20 billion dollars a year in Brazil. This volume could grow if more companies were concerned about reusing and recycling materials. However, only 5% of organisations operating in Brazil have some form of reverse logistics (RODRIGUES E WANDEREI, 2019).

Reverse logistics in the construction industry, although still in its infancy in Brazil, is already a reality in many countries, as it has proved to be an important tool for reducing environmental impacts, reducing the use of natural resources and reducing the volume of waste disposed of. In addition, it is an important mechanism for social, economic and sustainable development, since it benefits the emergence of new businesses, brings financial returns to the companies that adhere to it and preserves the environment for future generations.

Risks are present in all activities involving the construction industry and can jeopardise people's health and safety, as well as the company's profits, due to the waste of material that could be reused or recycled and returned to the production chain (SANTOS, 2013).

If applied to the construction industry, it could generate major benefits, reducing the consumption of raw materials through the recycling and reuse of materials, consequently reducing the volume of waste disposed of and the costs added to the sector, bringing improvements not only to companies, but to society as a whole (RIBEIRO, MOURA E PIROTA, 2016).

Logistics makes it possible to efficiently and effectively plan, implement and control the return or recovery of products to their production cycle, reducing the environmental problems caused by the generation of a high volume of waste and its improper disposal in the construction sector (RIBEIRO, MOURA and PIROTE, 2016).

The interest in eliminating waste in the construction industry is due to the growing concern for the environment and, above all, the concern to fulfil the wishes of suppliers and customers to reduce costs. "Companies are obliged to carry out studies into the disposal of materials, by means of environmental legislation, so that there is no degradation of the environment" (PINTO, 2013).

By reusing or recycling, companies create a differentiated image with new profit opportunities by introducing environmental concerns as a corporate strategy, and they are constantly looking for products and processes with less environmental impact and in line with sustainable development (LACERDA, 2014).

Recycling is a form of relationship between the companies involved in product return, as well as the management of recycling and the management of the movement of products from one point in the chain to another, with the aim of capturing value for appropriate disposal or as a form of strategic positioning in the face of competition (FARIA, 2013).

According to the non-governmental organisation *Wold Wide Found For Nature* - Brazil (WWF-Brazil, 2016), "Sustainable development actually suggests quality rather than quantity, with a reduction in the use of raw materials and products and an increase in reuse and recycling". And, according to Veiga (2015, p. 9), Sustainable Development (SD) is the ambition for humanity to meet its current needs without compromising the possibility of future generations.

According to Fernandes (2016), the utilisation of construction and demolition waste through the use of recycled aggregates allows for the partial replacement of natural resources that would previously have been taken from nature, with possible solutions for the applicability of the 3R's policy (reduce, reuse and recycle) (Table 5).

Table 5 - Waste Solution Scheme

WASTE SOLUTION SCHEME			
Reduce	**Reuse**	**Recycling**	**Contain**
Materials Production	Demolition materials	Resistant materials	Trenches
Overproduction	Aggregates	Quality materials	Sediment
			Leaking Waste

Source: Batista, Marcelo Lopes (2022).

According to CONAMA Resolution 307 of 2002, the reuse of waste must comply with the provisions of the standardising laws for construction waste, which must first be sorted to enable the waste to be classified, to enable correct storage and proper disposal, taking into account the safety of people and health guarantees (SILVA et al., 2015).

Civil Construction Waste (CCW) is defined by CONAMA Resolution 307/2002 and by Law 12.305/2010, which classifies it as class A waste. It has great potential for recycling and reuse and can be reused in a variety of ways, with the aim of minimising negative impacts on the environment, as well as reducing costs and waste (EVANGELISTA; BASTOS COSTA; ZANTA, 2010).

Reducing aims to reduce the consumption of goods and services. (RODRIGUEZ, 2017). Recycling aims to transform different types of materials into raw materials for other products (ADDIS, 2010) and reusing aims to extend the useful life of products, enabling financial gains (MANCINI, et al., 2012).

CCW recycling should be seen as something permanent and not temporary (ALMEIDA et al., 2020). Matuti and Santana (2019) emphasise the importance of considering the economic perspective of disposing of waste in landfills, which generates high costs, making recycling a more attractive option, as well as the reduction in the purchase of new inputs, since these will be replaced by waste, generating a reduction in raw material acquisition costs.

Sustainable Applications of RCC

RCC is of excellent quality to be used as a raw material for aggregates, which can be used in various construction processes, such as making bricks, precast blocks, kerbs, pavements, laying and coating mortar, base and sub-base layers, pavements, primary road coverings, concrete manufacture, drainage layers, among others (BRAGA; VEIGA, 2017).

According to ABRECON (2017), the use of recycled aggregates in paving (base, sub-base or primary coating), in the form of gravel or in mixtures of waste with soil, is the most common application and this is due to the following advantages:

- It is the form of recycling that requires the least use of technology, which means the process costs less;
- It allows all the mineral components of the rubble (bricks, mortar, ceramic materials, sand, stones, etc.) to be used together, without the need to separate any of them;
- Energy savings in the rubble grinding process (compared to its use in mortar), since by using it in concrete, part of the material remains in coarse granulometres;
- The possibility of utilising a greater proportion of the rubble produced, such as that from demolition and small construction projects that do not require investment in grinding equipment;
- Greater efficiency of the waste when added to saprolitic soils compared to the same addition made with gravel. While the addition of 20% recycled rubble to the saprolitic soil generates a 100% increase in CBR, with natural gravel additions the increase in CBR is only noticeable with dosages starting at 40%.

In the production of mortar for laying and coating, recycled aggregates must have the same granulometry as sand, the advantages of which are (ABRECON, 2017):

- Use of the waste at the place of generation, which eliminates transport costs;
- Pozzolanic effect of ground rubble;
- Reduction in cement and lime consumption;
- Gain in mortar compressive strength.

The rubble processed by recycling plants can be used as aggregate for non-structural concrete, replacing conventional aggregates (sand and gravel), the advantages of which are (BRAGA; VEIGA, 2017):

- Utilisation of all the mineral components of the rubble (bricks, mortar, ceramic materials, sand, stones, etc.), without the need to separate any of them;

- Energy savings in the rubble grinding process (compared to its use in mortar), since by using it in concrete, part of the material remains in coarse granulometres;

- Possibility of utilising a greater proportion of the rubble produced, such as that from demolition and small construction projects that do not support investment in grinding / crushing equipment;

- Possible improvements in concrete performance compared to conventional aggregates when low cement consumption is used.

Table 6 shows the possible ways of using recycled aggregates from RCC.

Table 5 - Summary of Applications Produced with Recycled Aggregates.

Product	Waste	Use
Recycled aggregate	Concrete	Gravelling roads; filling construction voids; filling plant trenches and reinforcing embankments (slopes).
Recycled sand	RCC	Aggregate for mortars; concrete with no structural function.
Recycled gravel	RCC	Base and sub-base or primary coating; concrete with no structural function.

Source: (Braga; Veiga, 2017) adapted by the author

Quality and Environmental Certifications in Construction

Companies in the construction industry in Brazil are currently showing great interest in obtaining quality and environmental certifications. Adhering to certifications brings about a continuous change in quality gains, an improvement in company organisation, a reduction in material waste, a reduction in rework, improvements in staff health and safety, as well as being used as a corporate strategy (QUALYTEAM, 2021).

The most widely adopted quality certifications are, in particular, the ISO 9001:2015 standard and the Brazilian Habitat Quality and Productivity Programme - PBQP-H (Figure 10).

Figure 8 - Differences between Standard 9001 and PBQP - H.

CERTIFICADO	ISO 9001	PBQPH
Significado	ISO, significa igualdade (origem grega)	Programa Brasileiro de Qualidade e Produtividade do Habitat
Aplicação	Em todas as empresas	Construção Civil
Abrangência	Padrão Internacional	Somente Brasil
Norma	NBR 9001*	SIAC –Sistema de Avaliação da Conformidade de Empresas e Serviços e Obras da Construção Civil
Processos Analisados	Os que forem importantes à empresa	A própria Norma, estabelece os processos que devem ser analisados
Processos Analisados	Os que forem importantes à empresa	A própria Norma, estabelece os processos que devem ser analisados
Versão	2015	2017
Valor	R$ 170,00	Gratuito (Site do Ministério das Cidades)

Source: Gonzalez, 2017, p. 20

One way to expand a quality management system would be to implement the SiAC of the Brazilian Programme for Quality and Productivity in Housing Construction (PBQP-H). Implementing a quality management system by obtaining certification under the PBQP-H SiAC is a process that is accessible to both small and large construction companies (MOREIRA, 2018). The PBQP-H aims to prepare the construction sector in favour of better quality, through labour, qualification of construction companies, material suppliers, among others (SILVA and PAINS, 2017).

According to Oliveira (2017), ISO 9001:2015 would be another target for construction companies, as it aims to indicate requirements for the quality management system of companies, bringing them improved image and credibility, customer satisfaction, better integration of processes and better

43

evidence for decision-making, creating a culture of continuous improvement and greater employee involvement.

According to Arruda (2017), ISO 9001 and ISO 14001 are the best-known standards, which deal with quality management systems and environmental management systems respectively. These standards work together to create a single system for managing quality in companies. Companies of any size can use the standard. The general aim of ISO 9001 is to improve and standardise organisations' processes, increasing the quality of products and services managed.

To obtain ISO certification, you don't have to pay any fees to the government, but it is a condition of the standard that the company is consistent with its legal and tax obligations. ISO is considered a seal of quality, in which various companies around the world affirm the quality of their products or services, expanding the market and gaining credibility and trust (MARTINS and DOMINGUES, 2018).

Vieira & Oliveira Neto (2019) emphasise that, given the various challenges facing the construction industry, it is essential to meet at least the basic quality rules for buildings, which are increasingly demanding productivity, competitiveness and sustainability due to the economic crisis and the demands of the consumer market. Hence the importance of adhering to quality and environmental certification programmes.

According to the Brazilian Association of Architecture Firms (2007, p.2), a development becomes sustainable when:

> (...) are linked to the issues of: internal and external environmental quality; reducing energy consumption; reducing waste; reducing water consumption; taking advantage of local natural conditions/planting and analysing the surroundings; recycling, reusing and reducing solid waste; and innovation.

It is necessary to study the feasibility of applying environmental certifications in the Brazilian context, since they guide companies interested in building more sustainably. Environmental certification has become extremely important in the construction industry, as it aims to significantly reduce the impacts generated. Certifying means directing activities and confirming the sustainability of a development, but also positively influencing construction companies to try to mitigate these impacts. Another advantage for entrepreneurs is to increase the possibility of sales by advertising the differential of having a certified building (MERTEN; VIAN, 2016).

There are different certification systems, including GBTool, Green Globes, BREEAM, HK-BEAM, Minnesotta Sustainable Design Guide (MSDG), CASBEE, AQUA, LEED, LiderA, Casa Azul, among others (Figure 11).

Figure 9 - Criteria Established for Obtaining Certifications.

Source: Own authorship

AQUA: This stands for High Environmental Quality. This certification was derived from the Démarche HQE certification, developed by the French association HQE. The Vanzolini Foundation is responsible for the AQUA process

45

in Brazil and the certification can be applied in all phases of a building's life, in various branches of civil construction and, for this reason, its auxiliary technical references have been divided into types: residential buildings, commercial buildings, schools, hotels and neighbourhoods and allotments. The certification's technical benchmark is structured around two instruments to assess the performance achieved: the Development Management System (SGE), to assess the environmental management system used by the developer; and the Building Environmental Quality (QAE), to assess the architectural and technical performance of the construction (FUNDAÇÃO VANZOLINI, 2013). The Vanzolini Foundation reference explains that the Environmental Quality of the building is structured into 14 categories, which can be grouped into four families: Site and Construction; Management; Comfort and Health. The biggest player in AQUA certification is the designer.

LEED: According to the 2009 U.S. Green Building Council (USGBC) reference, LEED NC version 3.0 for new construction, the Green Building Rating System is a system for certifying the sustainability and reduction of the environmental impact of buildings that adopt practices that take the environment into account. The LEED certification system is voluntary and market-orientated. Through analyses and guarantees of its environmental performance in the project's life cycle, sustainable practices are defined in the design, construction and operation phases. The system is organised into five categories: Sustainability of space; Rational use of water; Energy and atmosphere; Materials and resources; Indoor environmental quality.

LIDERA: This is a sustainable assessment system developed from studies carried out by the Department of Civil Engineering and Architecture at the Instituto Superior Técnico (IST) in Portugal in 2000. In 2005, version v1.02 was launched, followed by version v2.00 which, in addition to evaluating the building itself, encompasses the surrounding environment, integrating the building with the community. The principles governing LiderA are specified by Pinheiro (2011), divided into six strands and subdivided into 22 areas and 43 criteria (Figure 11): Sustainable Use; Socioeconomic Living; Environmental Comfort; Environmental Loads; Resources and Local Integration.

CASA AZUL: This certification is described in the CAIXA ECONÔMICA FEDERAL guide, Good Practices for More Sustainable Housing, from 2010. The guide is organised into two parts. The first presents the impacts of the construction production chain and the sector's needs in relation to sustainability. The second part is related to the challenges of the sustainable construction agenda, which structure the Seal. The Seal applies to all types of housing development projects submitted to CAIXA. Construction companies, public authorities, public housing companies, cooperatives, associations and organisations representing social movements can apply. The method used to award the seal consists of checking, during the technical feasibility analysis of the project, that it meets the criteria established by the instrument, which encourages the adoption of sustainability-orientated practices. The Seal has 53 assessment criteria, divided into six categories that guide the classification of projects: Urban Quality; Design and Comfort; Energy Efficiency; Conservation of Natural Resources; Water Management and Social Practices.

Chapter IX - Technological Innovations and the Construction Industry

BIM (Building Information Modelling) **technology**: The search for constant advances in construction is moving towards industrialisation as a way of optimising the process. In this scenario, BIM is proving to be a great tool, capable of detailing projects in a more assertive and clear manner, to improve planning and the processes to be carried out (NOBRE, 2021) (Figure 12).

Figure 10 - Modular Hospital.

* Modular hospital, built in 33 days, to serve COVID patients in 2020, using the BIM (Building Information Modelling) tool. Source: https://brasilaocubo.com/portfolio/hospital-mboi-mirim

Building Information Modelling (BIM) is a method of developing and using a model in a virtual environment to simulate the planning, construction, performance and operation of a building. The end result of the process is a model that is robust in information, intelligent and a parameterised digital representation of the building, which can be fed and used by various users (owners, architects, engineers, builders, subcontractors and suppliers), which can lead to an improvement in the decision-making process, making it more assertive and agile (ASSOCIATED GENERAL CONTRACTOR OF AMERICA, 2005).

According to Nobre (2021), when used in the planning phases of a project, BIM (Building Information Modelling) is a very versatile tool. By using the 3D BIM model and 4D simulations, it is possible to identify numerous uses, such as: surveying inputs, visualising the project, simulating sequencing to check for possible flaws in planning, reporting on the physical progress of the work and improving the flow of information on waste management and safety on construction sites, among others.

BIM technology can be used by the architecture, engineering and construction industry as a platform to minimise construction waste in their projects (LIU et al., 2011).

As well as being a technological innovation, BIM is also a significant change in the overall processes of the construction industry (DENG et al., 2019).

BIM-based building design offers construction professionals possibilities to carry out efficient planning and management in relation to the use of building materials. According to Cash (2019) and O'Reilly (2017), it is particularly during the design stages that the reduction of construction waste generation could be supported and improved through its use.

Won, Cheng and Lee (2016) quantified the construction waste caused by design errors that could be avoided using a BIM-based design validation process, which eliminates 4.3 per cent to 15.2 per cent of construction waste.

The sustainability of a construction site can be positively impacted by the use of the BIM tool, enabling both better waste management and the flow of information between environmental consultants and construction companies (JUPP, J. 2017; GUERRA, B. C.; LEITE, F.; FAUST, K. M. 2020).

Industrialised Building Systems: CDW is generated in three distinct stages: during construction, maintenance and demolition (SILVA, 2014). The construction industry stands out as a major generator of waste, and the amount of waste is directly proportional to the degree of development of a city, the result of greater economic activity and the resulting consumption habits, spaces for work, housing and leisure. All activities consume time and resources, but some add value to the product. When it comes to construction, laying bricks, for example, to make masonry, adds value, but rework and employees standing idle due to a lack of materials are examples of activities that do not add value, i.e. they are unproductive (FORMOSO, 2002).

One of the alternatives for this reduction would be the adoption of industrialised construction systems, whose main characteristics are greater planning, technical and economic feasibility studies and logistics, as well as better working conditions and environmental performance (LEAL et al., 2015).

Industrialisation in construction consists of carrying out several stages of a building at the same time, making it a simultaneous process (DONIAK, 2012).

The benefits of industrialisation are many: control and safety facilities; a high level of quality in production; a low level of waste; a reduction in the number of workers; modulation, uniformity and standardisation; speed of execution and, above all, a reduction in the generation of waste, making it the future of construction (MOURA and SÁ, 2013).

The industrialised construction system is made up of a set of rules and coordinated parts, in which the construction details already have a solution, before they are designed and applied to a project, according to Richard (2012), the products are the construction systems and not the buildings. Chirinéa (2018) adds that in Brazil, the construction scenario is far removed from this concept, since the solutions for the project are defined in the field.

According to the authors NUNES E SOUZA (2017), industrialised construction systems are presented as alternatives to the traditional construction method and some examples are: steel as an alternative for the structure, Light Steel Framing (LSF) and Steel Deck as alternatives for slabs and Light Steel Framing and Drywall as alternatives for sealing systems. These systems have the potential to reduce waste in construction, as well as contributing to productivity gains, rationalisation and modernisation (Figure 13).

Figure 11 - Building under construction using the Light Steel Framing (LSF) industrialised construction system.

Source: Inova Steel (2017)

Concrete, for example, widely used in conventional construction, is one of the materials that contributes most to greenhouse gas emissions on the planet. Replacing this material with steel structures of the Light Steel Frame type implies a significant reduction in these indices, as well as making a huge contribution to sustainability (Figure 14) (COSTA, 2019).

Figure 12 - Reducing Environmental Impacts in Sustainable Buildings.

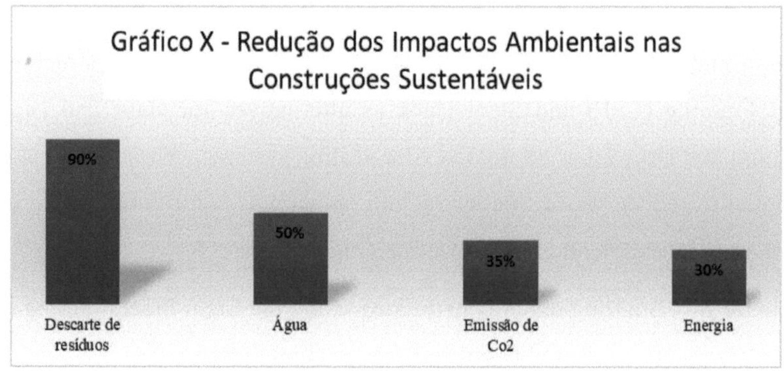

Source: Kwai (2013)

According to Lessa (2005), the drywall sealing system has numerous advantages over conventional masonry due to its high versatility, acoustic performance when coupled with mineral wool and double sheeting, flat surfaces, rationalised solutions for other subsystems such as electrical and plumbing, high productivity, time optimisation, greater quality control and, above all, the reduction in project loads due to lighter materials than conventional masonry.

> The current market offers three types of plasterboard, each adapted to a specific area: drywall, known as standard board (ST); moisture-resistant board (RU), known as green board, for wet areas; and fire-resistant board, with a reddish colour, used for areas with special fire-resistance requirements (FREITAS & CASTRO, 2006).

According to MEDEIROS *et al.* (2014), the main reasons for using Light Steel Framing (LSF) are: the use of industrialised components and rational assembly, allowing for a reduction in the delivery time of the work, making the system more compatible with steel structures which also have these potential properties; considerable reduction in loads when compared to other types of sealing; weight less than 1/3 of the weight of ceramic block masonry and weight less than 1/4 of concrete block masonry; probable quality control of the execution of the service provided by the separation of the stages (assembly of the structure, sealing with panels and final finishing); Ease of assembly, handling and transport, due to the simplicity of the production chain.

> Due to its speed of execution and rationalised use of materials, LSF is expected to gain more and more ground in the domestic construction market. It is possible to reduce the final schedule of some projects, with a direct impact on their fixed cost, as has been the case in several projects carried out in Brazil (MEDEIROS et al. 2014).

The metal structure has advantages over other materials in terms of time, because the execution time is short, due to the possibility of having simultaneous processes, i.e. while the foundation is being made, the metal parts can already begin production; material waste is relatively low, and there is no need to handle cement, sand, wood, gravel or reinforcement on a large scale. In addition, a metal skeleton weighs on average ten times less than a concrete skeleton (BELLEI et al., 2004).

According to LEAL *et al.* (2015), the use of industrialised construction methods ensures that civil construction has an accelerated rate of production compared to traditional systems. This is due to the fact that these systems are prefabricated and only have to be assembled on the building site. This gain in speed can be seen in the assembly of pillar structures, beams and slabs and drywall, which can reduce execution time by around 40 per cent compared to conventional methods, due to the accumulation of simultaneous stages that this system provides.

According to Nunes and Souza (2017), a comparison between two construction projects using different building systems revealed the following:

- The conventional construction system, Residencial Majestic, in Tubarão/SC, with 28 floors totalling 9,324.35m², was carried out over 6 years;

- An industrialised construction system, Tower E of the W Torre Complex, in the south of São Paulo, with 26 floors and a total built area of 23,600m², was completed in just 11 months, between 2012 and 2013.

According to Farias; Medeiros and Freitas (2015), with regard to the relationship between construction and technological innovations, there is a need to promote and create environmental regulations to enable environmental management, promoting investment incentives and technological innovation, in order to speed up production and reduce the waste of raw materials.

In Brazil, the construction sector still sees the industrialised building system as a recent technological innovation. Factors such as: lack of mastery of the construction method by design professionals; labour with a low level of education and no specific training; cultural issues, among others, have still been barriers to LSF being adopted more widely in Brazil's construction market (SOUZA, 2014).

Chapter X - Solid Waste Management Plans

Law 12.305/2010 established Solid Waste Plans as legal instruments for solid waste management, determining that they can be organised according to the different legal competences of the different federal entities, and that they should be widely publicised in terms of their content, as well as subject to social control in their formulation, implementation and operation (ANTUNES, 2014). These plans are divided into four levels, and each sphere of the country is responsible for drawing them up (BRASIL, 2010):

- **The Federal Government, through the Ministry of <u>the</u> Environment:** responsible for the <u>National Solid Waste Plan,</u> which is in force for an indefinite period and has a twenty-year horizon, and must be updated every four years.

- **States:** responsible for the <u>State Solid Waste Plans,</u> as a condition for obtaining funds from the Federal Government to finance projects aimed at solid waste management, also valid for an indefinite period and with a twenty-year horizon, and which must be updated every four years. In addition, states will also be able to draw up micro-regional Solid Waste Plans and Solid Waste Plans for Metropolitan Regions or Urban Agglomerations.

- **Federal District and Municipalities:** responsible for <u>Municipal Integrated Solid Waste Management Plans and Intermunicipal Solid Waste Plans,</u> the latter when more than one municipality is involved. These are conditions for municipalities and the Federal District to obtain funds from the Federal Government to finance services or undertakings related to urban cleaning and solid waste management.

- **Construction companies, generators of waste from public sanitation services, industrial waste, waste from health services, agroforestry**

waste, waste from transport services, mining waste and commercial establishments or service providers that generate hazardous waste: responsible for the <u>Solid Waste Management Plan,</u> to be drawn up in accordance with regulations or standards established by the National Environment System - SISNAMA.

Figure 15 shows a flowchart demonstrating the hierarchical structure between the various Plans and those responsible for them:

Figure 13 - Flowchart of the Hierarchical Structure of Solid Waste Plans

Source: Adapted from Brum (2013).

After 12 years of waiting, the National Solid Waste Plan (Planares) was instituted by Federal Decree No. 11,043 of 13 April 2022, and is now valid throughout the country. Drawn up through a Technical Cooperation Agreement between the Ministry of the Environment and the Brazilian Association of Public Cleaning and Special Waste Companies (ABRELPE), Planares establishes guidelines, strategies, actions and targets to modernise solid waste management in the country, in order to put into practice the objectives set out in the National Solid Waste Policy - Law No. 12.305 of 2010 (MMA, 2022). Article 3 states that state, micro-regional, metropolitan region or urban agglomeration, intermunicipal and municipal solid waste plans must comply with the National Solid Waste Policy and the National Solid Waste Plan.

The National Solid Waste Plan (Planares), in addition to setting targets for compliance with various points of the law, calls for an increase in waste recovery, setting a target of 50 per cent utilisation in 20 years. This means that half of the waste generated will be recovered through recycling, composting, biodigestion and energy recovery, which is a major step forward compared to the current scenario in which only 3% of solid urban waste is recovered. The plan also provides for an increase in the recycling of construction waste, encourages the recycling of materials, contributes to the creation of green jobs, as well as making it possible to better fulfil international commitments and multilateral agreements, with clear indications for the reduction of greenhouse gas emissions (MMA, 2022).

> "A very important step has been taken with the approval of the National Solid Waste Plan, the main national planning document for solid waste management, which has now become a reality. The plan presents possible ways to achieve the objectives outlined in the National Solid Waste Policy, with concrete actions to obtain tangible results that translate into better health and quality of life for Brazilians" (FRANÇA, ANDRÉ, 2022).

It is worth noting that the absence of a Municipal Integrated Solid Waste Management Plan does not exempt private institutions, such as companies in the construction sector, from responsibility for drawing up, implementing and operationalising the Solid Waste Management Plan (BRASIL, 2010).

Chapter XI - Construction Waste Management Plans

Conama Resolution 307 of 5 July 2002 established guidelines, criteria and procedures for the correct management of construction waste, regulating essential actions in order to reduce environmental impacts. This Resolution underwent three revisions, the first on 16 August 2004 (Resolution 348/2004), the second on 24 May 2011 (Resolution 431/2011) and the third on 19 January 2012 (Resolution 448/2012), adapting to Law 12.305/2010. In this revision, the Integrated Civil Construction Waste Management Plan (PIGRCC) and the Municipal Civil Construction Waste Management Programme (PMGRCC), which were the responsibility of the municipalities, as previously provided for in Article 5 of Resolution 307/2002, had their structure altered and unified, now called the Municipal Civil Construction Waste Management Plan.

According to Article 5 of CONAMA Resolution 448 of 18 January 2012, the Municipal Construction Waste Management Plan is the instrument for implementing construction waste management and must be drawn up by municipalities and the Federal District in line with the Municipal Integrated Solid Waste Management Plan.

They must be included in the Municipal Construction Waste Management Plan:

> I - The technical guidelines and procedures for exercising the responsibilities of small generators, in accordance with the technical criteria of the local urban cleaning system and for the Construction Waste Management Plans to be drawn up by large generators, making it possible for all generators to exercise their responsibilities (art. 6 of CONAMA Resolution 448/2012).
>
> II - The registration of areas, public or private, suitable for receiving, sorting and temporarily storing small volumes, in accordance with the size of the municipal urban area, enabling the subsequent destination of waste from small generators to processing areas;

III - The establishment of licensing processes for waste processing and storage areas and for the final disposal of tailings; IV - The prohibition of the disposal of construction waste in unlicensed areas;

V - Encouraging the reinsertion of reusable or recycled waste into the production cycle;

VI - The definition of criteria for the registration of hauliers;

VII - Guidance, inspection and control of the agents involved;

VIII - Educational activities aimed at reducing the generation of waste and enabling it to be segregated (BRASIL, 2002).

Article 2 of CONAMA Resolution 307/2002 establishes generators as individuals or legal entities, public or private, responsible for activities or undertakings that generate the waste defined in the resolution (BRASIL, 2002).

Pozzobon (2014, p. 72-89), states that:

In turn, construction waste is the result of the growth and development of cities, the increase in population and the modernisation and expansion of roads and buildings that make up the so-called 'artificial' (urban) environment. Construction waste is therefore a permanent reality - not a passing one - and precisely for this reason it deserves specific treatment, with its own legislation and regulations, as has been the case in Brazil.

Construction Waste Management Plans will be drawn up and implemented by large generators and will aim to establish the necessary procedures for the environmentally appropriate handling and disposal of waste, according to Article 8 of Resolution 448/2012.

Art. 9 of CONAMA Resolution 448/2012 establishes that Construction Waste Management Plans must include the following stages: I - characterisation: in this stage the generator must identify and quantify the waste; II - sorting: this must preferably be carried out by the generator at source, or be carried out in licensed disposal areas for this purpose, respecting the waste classes established in Art. 3 of this Resolution; III - packaging: the generator must ensure that the waste is contained after generation until it is transported. III - packaging: the generator must ensure that the waste is contained after generation, until the transport stage, ensuring in all cases where possible, the conditions for reuse and recycling; IV - transport: must be carried out in accordance with the previous stages and in accordance with the technical standards in force for waste transport; V - destination: must be provided for in accordance with the provisions of this Resolution (BRASIL, 2002).

Chapter XII - Overview of Construction Waste in Brazil

According to ABRELPE's Panorama (2022, base year 2021), more than 48 million tonnes of CDW were collected by Brazilian municipalities in 2021, which represents an increase of 2.9% compared to 2020. The amount collected per inhabitant was around 227kg per year and much of it was construction and demolition waste abandoned on public roads and streets. Just over half of the CDW collected in Brazil comes from the Southeast region (52 per cent) (Table 7).

Table 7 - Construction Waste Collected in 2021.

Region	Collected (tonnes/year)	Collection rate (kg/inhab/year)
North	1.870.260	98,90
North-East	9.481.605	164,40
Centre-West	5.403.095	323,40
South East	25.047.395	279,40
South	6.572.920	216,20
Brazil	48.375.275	**226,80**

Source: ABRELPE, 2022 (Base Year 2021).

According to Costa (2014), the recycling of CDW contributes to extending the useful life of landfills, especially in large cities where construction is intense and there is a shortage of landfill space.

In Brazil, the proper disposal and recycling of rubble is still not widespread (OLIVEIRA and MENDES, 2008).

According to the National Solid Waste Plan (Planares, 2022), the following types of management units were identified in Brazil in 2018:

a. RCC Recycling Area (or Rubble Recycling Unit): units dedicated to transforming RCC into other materials for reinsertion into the construction industry;

b. RCC and bulky materials transshipment and sorting area (ATT): units dedicated to storing and sorting RCC, for subsequent transfer to other units (for final disposal or processing);

c. RCC landfill (or inert landfill): a site for the final disposal of RCC, especially after it has been sorted;

d. Transhipment units: units dedicated to temporary storage, for subsequent transfer to other units (for sorting, processing or final disposal);

e. Sorting Units (or Sorting Sheds or Plants): units dedicated to sorting RCC;

f. Other.

Due to the predominance of the construction pattern in Brazil, the highest percentage of material found in CCW is mortar, especially concrete mortar, which is used in the composition of structures (PLANARES, 2022).

Waste characterisation is the start of the management process for waste produced on site and must be classified in accordance with specific regulations (NBR 10.004/2004 and CONAMA Resolution 307/2002).

Table 8 presents data showing that in the northern region, only the states of Pará and Tocantins have management units.

Table 8 - Management Units in the Northern Region.

Region/UF	RCC recycling area	RCC and bulky materials transhipment and sorting area	RCC landfill (inert)	Total
AC	0	0	0	0
AM	**0**	**0**	**0**	**0**
AP	0	0	0	0

PA	0	0	1	1
RO	0	0	0	0
RR	0	0	0	0
TO	0	1	0	1
Total Northern Region	0	1	1	2

Source: SNIS-RS, 2019 (base year 2018).

The Southeast region, especially the states of São Paulo and Minas Gerais, is home to the largest number of CHW management units in the country (Table 9).

Table 9 - Management Units in the Southeast Region.

Region/UF	RCC recycling area	RCC and bulky materials transhipment and sorting area	RCC landfill (inert)	Total
ES	0	0	1	1
MG	4	2	17	23
RJ	1	1	1	3
SP	16	17	22	55
Total Southeast Region	21	20	41	82

Source: SNIS-RS, 2019 (base year 2018).

According to the Panorama of Solid Waste in Brazil 2018/2019 (ABRELPE, 2019), the percentage of construction waste recycling was calculated for the country's regions based on the proportion of CHW collected in each region, and the intermediate values for the country and the regions were obtained through linear interpolation, considering the current value and the final target value. Table 4 shows the percentage of construction waste recycled between 2020 and 2040, according to each region, with a target of 25 per cent to be achieved in that year.

Table 10 - Projection of MSW recycling up to 2040.

Region	2020	2024	2028	2032	2036	2040
North	0,27%	0,41%	0,55%	0,69%	0,83%	0,96%
North-East	1,40%	2,11%	2,82%	3,52%	4,23%	4,94%
Centre-West	0,77%	1,16%	1,55%	1,94%	2,33%	2,72%
South East	3,68%	5,56%	7,43%	9,30%	11,17%	13,05%
South	0,94%	1,42%	1,90%	2,37%	2,85%	3,33%
Brazil	**7,06%**	**10,65%**	**14,24%**	**17,82%**	**21,41%**	**25,00%**

Source: Planares, 2022

Chapter XIII - Overview of the current construction and demolition waste management scenario in the city of Manaus.

Manaus is bordered to the north by the municipality of Presidente Figueiredo; to the south by the municipalities of Careiro da Várzea and Iranduba; to the east by the municipalities of Rio Preto da Eva and Itacoatiara; and to the west by the municipality of Novo Airão (Figure 16). According to the IBGE (2021), the municipality of Manaus has a territorial area of 11,401.09 km², a population of 2,255.90 inhabitants and showed a population growth of 20.12% in 2021, compared to 2010. Driven by this growth, the construction industry has grown stronger and the generation of construction waste has increased considerably (OLIVEIRA, M.P.S. LÂMEGO, 2021).

Figure 14 - Geographical boundaries of Manaus.

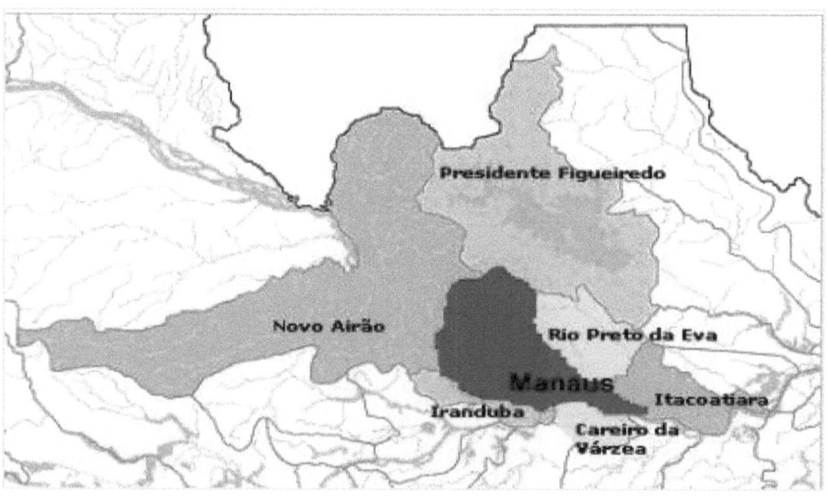

Source: PDRSM/2010

According to Klaus (2014), in the last twenty years, the solid waste generated in Brazil has changed greatly in terms of quantity, quality and composition. This is due to the rapid and disorganised growth of cities and, at the same time, changes in citizens' habits, which has generated waste that is very different from what cities produced years ago.

Waste Management Plan and Environmental Legislation: On 9 November 2011, the Manaus Solid Waste Master Plan (PDRS) was approved, published by Decree No. 1.349/2011, taking into account Federal Law No. 11.445/2007 and Law No. 12.305/2010, covering technical, environmental, economic and social issues such as improving the waste collection and treatment infrastructure network, reducing the generation of solid waste and encouraging reuse, recovery and recycling, promoting the economic sustainability of the waste management model, formalising, training, professionalising and fully integrating the informal sector in waste management. The Manaus Solid Waste Master Plan was structured for a 20-year time horizon, i.e. from 2009 to 2029.

On 20 January 2010, the Municipality of Manaus enacted Municipal Law 1.411, which instituted the urban cleaning system of the Municipality of Manaus at the time (SLUMM). However, this law only came into force on 1 January 2011.

In November 2015, a proposal for the Manaus Municipal Plan for Integrated Solid Waste Management (PMGIRS) was submitted to a public hearing at the Manaus City Council, in compliance with the 4-year update period set out in the PNRS.

In line with the National Solid Waste Policy, the State Solid Waste Policy (PERS/AM) was established by Law No. 4,457 of 12 April 2017. On 30 January 2020, Decree No. 41,863/2020 was published, which provides for the State Solid Waste Policy and regulates provisions of Laws No. 4,457 of 12 April 2017, Law No. 4021 of 12 April 2014 and promulgated Law No. 249 of 31 March 2015 and makes other provisions, such as the creation of the State Solid Waste Committee

(CERS) and its attributions, the State Solid Waste Information System (SEIRES) and the Certification of Sustainable Practices.

Solid Waste Management: Article 10 of Law 12.310/2010 states that the Federal District and Municipalities are responsible for the integrated management of solid waste generated in their respective territories, without prejudice to the control and inspection powers of the federal and state bodies of SISNAMA (National Environment System), SNVS (National Health Surveillance System) and SUASA (Unified Agricultural Health Care System), as well as the generator's responsibility for waste management, in accordance with the provisions of this Law.

In Manaus, the Municipal Department of Urban Cleaning (SEMULSP) is the department responsible for public cleaning and solid waste management services, and therefore has the legal authority to organise and administer them. It was created by Law No. 761, of 4 May 2004, and the last legislative change to its structure was defined by Decree No. 2,583, of 23 October 2013. The Secretariat receives solid waste from the cities of Manaus, Careiro da Várzea and Iranduba. According to the Metropolitan Region's Solid Waste and Selective Collection Plan (PRSCS-RMM, 2017), the other municipalities in the interior of Amazonas dispose of their solid waste in landfills, vacant lots or municipal rubbish dumps.

> However, this reality has not been extended to the municipalities of Amazonas, most of which have open-air rubbish dumps, generating various environmental and public health problems, sometimes jeopardising much of the soil and drinking water sources (JACINTO, ANA CAROLINA, 2016).

Public cleaning services are carried out directly by the Secretariat and also by a company contracted through a public tender. The planning, regulation and inspection of these services are the responsibility of the Sub-Secretariat for Operations (SubOp), a body linked to SEMULSP.

Article 133(I) and (II) of Law 1.411/2010 of the Municipality of Manaus considers large waste generators to be public, institutional, service, commercial and industrial establishments, among others, that generate waste classified by NBR 10004 as Class II, in a volume of more than 200 litres per day, as well as establishments that generate CDW, in a volume of more than 50kg per day.

Large generators (Table 11) and their transporters must be registered with the municipal body responsible for managing the services, declaring the monthly volume and mass of waste produced, the operator contracted for collection, transport and the intended final disposal method (PDRSM, 2010).

Table 11 - Generator Groups and their Limits.

GROUP	LIMIT
Small Solid Waste Generator	Those who generate up to 200 litres or 100kg of household waste per day.
Large Generator of Solid Waste	Those who generate more household waste than this limit.
Small Generator of Construction and Demolition Waste	Those who generate a volume of up to $1m^3$ of inert waste.

Source: PDRSM (2010)

According to the PDRSM, construction and demolition waste must be packaged in resistant plastic bags with a minimum capacity of 20 litres (nominal bag capacity of up to 20kg). Large generators must package their waste in accordance with the Municipal Integrated Construction Waste Management Plan drawn up by the municipality. For small volumes, the responsibility lies with the municipal government, and for large volumes, the responsibility lies with the private sector, under the supervision of the government, which will have to define, licence and make available management areas for CCW.

The specific objectives of the PDRSM are: Preventing waste generation at source; Reducing solid waste generation; Encouraging reuse, recovery and recycling; Preventing and correcting environmental impacts; Promoting the economic sustainability of the waste management model, among others. The results were designed to be achieved in stages, divided into various short-, medium- and long-term goals, over the period from 2010 to 2029;

Responsibility for waste lies with the generator, with the other participants in the chain of activities being jointly and severally liable, within the scope of their participation, and the public authorities having the role of disciplining and supervising the activities of private agents.

The CONAMA Resolution makes it compulsory for municipalities to implement an Integrated Construction Waste Management Plan. The Municipal CDW Management Programme must be drawn up, containing the technical guidelines and procedures for exercising the responsibilities of small and large generators and transporters, and CDW management areas must be defined, licensed and made available.

Large generators must package their waste in accordance with the Integrated Construction Waste Management Plan drawn up by the municipality, and large generators must also be identified and registered, and hauliers accredited;

Important points of PERSCS-RMM:

- ✓ The main objectives of the PRSCS-RMM were to carry out a diagnosis of Solid Waste Management in the municipalities of the Metropolitan Region of Manaus (RMM), as well as to propose actions aimed at better management of this issue and the financial sustainability of the System;

- ✓ According to the PRSCS-RMM (2017), in the municipality of Manaus there is still no organised system for the collection, treatment, disposal or recycling of construction and demolition waste (CDW);

✓ As a target, the Plan provides for the installation of PEV's, Transshipment and Sorting Areas (ATT), in compliance with CONAMA Resolution 307/02 and its amendments, considering a 20-year horizon.

> PEV - Voluntary drop-off point: Ecopoints for the temporary accumulation of construction and demolition waste, bulky waste, selective collection and waste with reverse logistics.

From Collection to Disposal: Construction waste is stored and collected by rubble companies (CRUZ, 2018) and in 2017 Manaus had 44 rubble companies (PRSCS-RMM, 2017). Waste hauliers call all waste from concrete, mortar, plaster, lime, ceramic materials, glass, concrete blocks, cellular concrete, clay bricks and soil inert (BRITO, 2015).

Table 12 shows the percentage distribution of construction waste generation in the city of Manaus in 2015, according to data from ATEMA (Manaus Rubble Workers Association) (BATISTA, 2022).

Table 12 - Waste Generated in the City of Manaus in 2015.

Waste	Percentage
Inert	64%
Wood/shoring	11%
Others	14%

Source: ATEMA (2015) - adapted

Wood and plastics are categorised according to their quantity and economic value. For example, 50 per cent of wood is sold to potteries as fuel for burning bricks, and the remaining 50 per cent of wood is sold for making furniture and boxes used in packaging (ATEMA, 2015).

In 2017, according to data provided by SEMULSP, the amount of RCC disposed of at the Manaus landfill was 2,070.10 tonnes between January and October 2017. The inert material collected and transported comes from companies providing services, such as rubble disposal, construction companies, industries, among others, which request authorisation to dispose of waste at the Manaus landfill (Figure 17).

Figure 1715 - Quantity of RCC sent to the Manaus Landfill (2017).

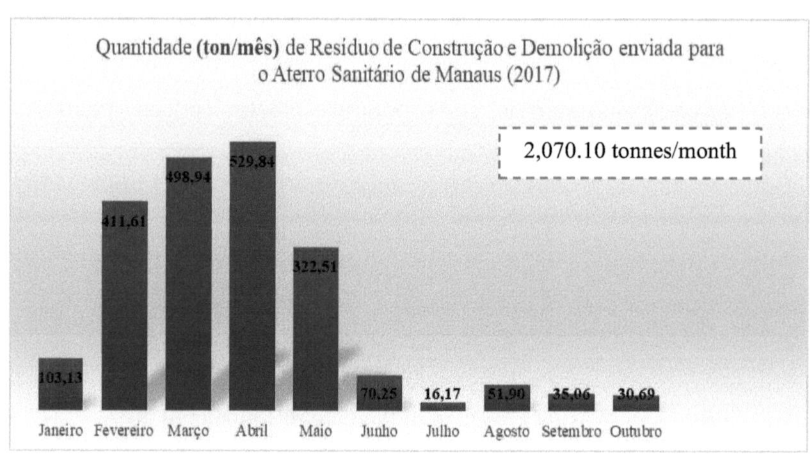

Source: SEMULSP (2017)

Irregular disposal of RCC: In 2018, 49 irregular disposal points were identified in the Distrito Industrial II neighbourhood, where 31 of the points are within the neighbourhood boundaries and the other 18 points are distributed along the neighbourhood's internal roads, where vacant lots, watercourses and branches are located. Green areas are the offenders' favourite places, and this is just a sample of how many sites are used as rubble dumps within the urban perimeter of the municipality of Manaus. The presence of this waste, in the open and in unsuitable areas, is the result of disorganised demographic expansion (TAVARES et al., 2019).

Figure 16 - Points of irregular deposits of RCC in the neighbourhood of Distrito Industrial II - Manaus-AM.

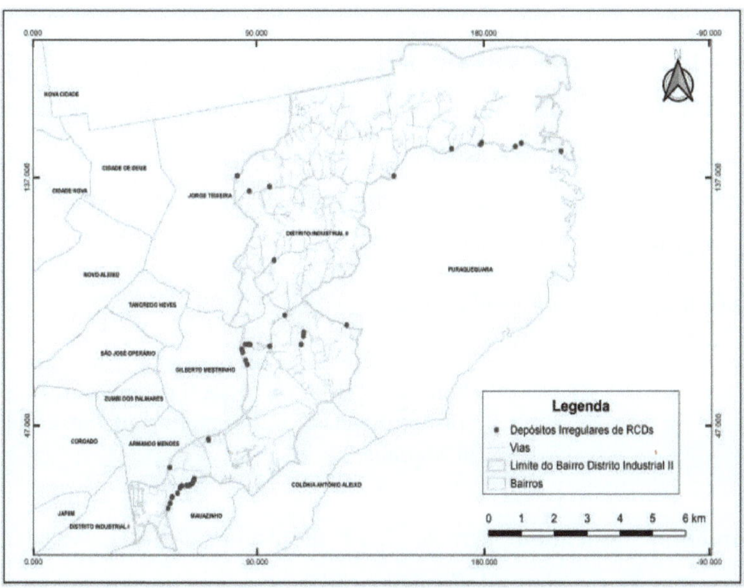

Source: (TAVARES et al., 2019)

"Construction waste may not be disposed of in urban solid waste landfills, in "no-go" areas, on slopes, bodies of water, vacant lots or in areas protected by law" (Art. 4 of Resolution No. 448 of 18 January 2012);

In Article 10, construction waste, after sorting, must be disposed of in the following ways:

I - Class A: Must be reused or recycled in the form of aggregates or sent to the Class A waste landfill to preserve the material for future use (Resolution No. 448 of 18 January 2012);

Sustainability: The public landfill in Manaus does not have a recycling and reuse plant for RCC and, therefore, there is no specific management of the reuse of this waste from a sustainability perspective (OLIVEIRA, 2021).

In order to carry out specific, large-scale works, such as PROSAMIM and the demolition of the Vivaldo Lima stadium for the construction of the Arena da Amazônia in 2014, there was a different approach to the issue of RCC. For PROSAMIM, two inert landfills were licensed in Manaus. For the demolition of the Vivaldo Lima stadium, a crusher/classifier was installed for the RCC from the project. However, at the end of the works, the equipment was removed and the systems deactivated (PRSCS-RMM, 2017).

According to data collected by Diniz (2018):

> "Despite this large amount of construction waste, there are still no effective measures for the correct disposal and recycling of these materials in Manaus."

Environmental Education: Article 7 of the National Solid Waste Policy establishes that Environmental Education (EE) is an instrument of Integrated Solid Waste Management. Federal Decree 7.404/10, in its article 77, defines the measures related to Environmental Education to be adopted by the Public Authorities for the implementation of the PNRS.

Environmental education is guaranteed by the 1998 Constitution of the Federative Republic of Brazil. Article 225 states that it is the responsibility of the Government to "promote environmental education, at all levels of education, and public awareness for the preservation of the environment".

The Amazonas State Solid Waste Plan (PERS-AM, 2017) highlights the Environmental Education process as an important action, but for it to be successful, society as a whole must be involved and participate in bringing about a change in mentality and, consequently, habits, in relation to the current problem. This awareness will not happen instantly, but through constant environmental education activity that guarantees the involvement and participation of everyone at school, in the family, in communities and in the state.

> "The practice of Environmental Education does not eliminate environmental problems, but it is an indispensable condition for doing so. The great importance of Environmental Education is to add knowledge, contributing to the formation of citizens aware of their role in preserving the environment, enabling them to decide on the environmental issues necessary for the development of a sustainable society" (PERS-AM, 2017).

Chapter XIV - The Current Scenario of Construction and Demolition Waste Management in the City of Manaus - Public Administration Bodies

The following information was collected through technical visits to the Municipal Secretariat for the Environment and Sustainability (SEMMAS), the Amazonas State Environmental Protection Institute (IPAAM) and the Municipal Secretariat for Public Cleaning (SEMULSP).

Municipal Plan for the Integrated Management of Solid Waste: According **to** the information collected during a visit to SEMMAS, with the Secretary of the agency, the Manaus Municipal Plan for the Integrated Management of Solid Waste (PMGIRS), which contains the topic of RCC, was updated in 2015 by SEMULSP. However, it was submitted for public consultation and was not approved, due to the recommendation of the Amazonas Public Prosecutor's Office to revise it in order to compose and eliminate some of the items specified, in terms of possible inconsistencies and improprieties, after which the necessary public hearings would be held to guarantee effective popular participation. In order to approve this Plan, it is necessary to comply with RECOMMENDATION No. 071/2015-MP-RMAM, issued by the Environmental Coordination Office of the Amazonas Public Prosecutor's Office, as well as updating it for the year 2023. According to the Secretary of SEMMAS, this update is being prepared under the responsibility of the Manaus Infrastructure Secretariat (SEMINF) and SEMMAS.

The Disposal of Solid Waste **from a Sustainability Perspective:** The Manaus Solid Waste Landfill is the city's main complex for the final disposal of solid urban waste, with an estimated area of 66 hectares, located at km 19 of the AM-010 motorway (Figure 19).

Figure 19 - Manaus landfill (2023).

Photo: Own Authorship (April/2023)

The complex is operated by two concessionaires and receives household waste, waste from mechanical removal, manual removal, pruning and selective collection and from third parties, i.e. companies providing services such as rubble disposal, construction companies, industries, among others, which request authorisation to dispose of waste. According to the SEMULSP report, in 2021 the landfill received 824,586 tonnes of waste collected by the two concessionaires and 13,700 tonnes deposited by third parties.

To enter the landfill, companies must have up-to-date registration with the Department and an environmental licence to carry out their activities. At the entrance to the landfill there is a scale, with a capacity of 60 tonnes, to check the weight of the trucks that transport the waste for disposal.

According to information provided by the director of the landfill during a visit to the complex, there are currently thirty-three third-party companies that have authorisation to dispose of waste materials at the site. According to the data provided by the Secretariat, the data in Table 7 and Figure 20 correspond to the estimated quantity of CCW destined for the landfill in 2022 (SEMULSP, 2023).

Table 1 - Estimated RCC (t/month) destined for the Manaus Landfill - 2022.

Jan	Feb.	Sea	Apr.	May	Jun.	Jul	Aug	Set	Out	Nov	Ten
364	400	437	419	564	640	725	944	1.082	1.360	1.548	1.745

Source: SEMULSP (April/2023)

Figure 170 - Estimate of RCC destined for the Manaus Landfill / 2022.

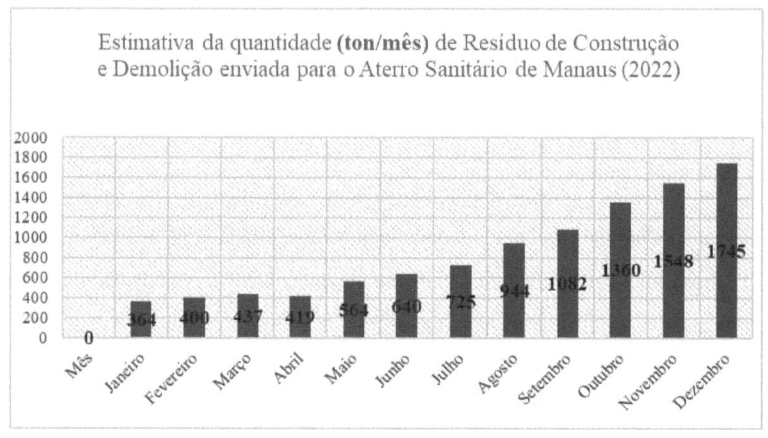

Source: SEMULSP (April/2023)

The waste received by the complex has three types of destination:

I) Grounding;

II) Recycling (sent to waste pickers' centres);

III) Production of organic compost, carried out by the Composting Plant located at the Landfill (SEMULSP 2021).

The Secretariat said that the landfill does not have an area for sorting, segregating and recycling RCC, that there is no recycling programme for type A RCC and that only this complex is licensed to receive this type of waste. Therefore, landfill is the destination of this waste.

The Environmental Protection Institute of the State of Amazonas (IPAAM) is the body responsible for issuing environmental licences to companies that collect, transport and dispose of RCC, as well as areas for the disposal of construction debris, and for monitoring and supervising the final disposal and/or processing of waste. Monitoring is carried out by following up on full compliance with the restrictions and conditions on the back of environmental licences. Illegal waste disposal sites are only inspected following complaints, according to information from the agency's Technical Management, i.e. there are no routine inspections due to the number of technicians, which is disproportionate to the city of Manaus.

According to IPAAM, areas are licensed by the Institute for the purpose of receiving rubble from specific construction sites. There are no other areas licensed by the agency to receive various types of rubble. For this reason, it is very common to find this waste deposited in inappropriate places, such as clandestine dumps, on the banks of streams and in vacant lots far from the city (figure 21). As described above, this leads to the silting up of the banks of watercourses, as well as the clogging of culverts and galleries (during floods), flooding and landslides, which have been occurring in the city very frequently, causing environmental, economic and social problems (SILVA et al., 2017).

Figure 181 - RCC deposited in an inappropriate place (Manaus/AM).

Source: Own Authorship (April/2023)

Environmental Education: Data collected from the Report (SEMULSP, 2021), informs that awareness-raising and environmental education actions related to waste are carried out by the Special Commission for Dissemination and Guidance of the Public Cleaning Policy (Cedolp) of Manaus City Hall. However, there are no figures to prove that this is actually happening, nor are there any practical results. With regard to RCC, there is no information about any work on the subject.

Chapter XV - *On-site* verification of construction and demolition waste management practices adopted on building sites in the city of Manaus

Through SINDUCON-AM, it was possible to access the list of forty companies that are members of the organisation and thus make it possible to contact them to schedule the technical visits.

Technical visits were made to the construction sites of companies A, B and C in the city of Manaus to check on the *spot* which RCC management practices have been adopted, taking into account the relevant legislation.

According to ABNT (NBR 12284, 1991, p.1), the construction site is defined as: "Areas intended for the execution and support of work in the construction industry, divided into operational areas and living areas". When drawing up a construction site, it is necessary to comply with the guidelines already laid down in Regulatory Standard NR-18, the aim of which is to implement safety control and prevention measures in construction processes, working conditions and the environment, as well as complying with the PGRCC, drawn up by large generators, in accordance with the relevant legislation.

> Waste Management Plans: In Art.8 - Construction Waste Management Plans will be drawn up and implemented by large generators and will aim to establish the necessary procedures for the environmentally appropriate handling and disposal of waste.

Company A - Conventional Masonry

Object: Two-storey residential building.

Location: Condomínio dos Pássaros - Tarumã.

The work is being carried out using traditional technology, i.e. conventional masonry, which, as previously established, generates a lot of waste due to all the situations already explained.

During a visit to company A's construction site, the observations described below were made:

It was found that the RCC is placed in front of the construction site, piled up next to the public pavement, without due concern for its correct packaging. Waste can be seen piling up without any kind of characterisation, sorting or segregation, mixed with raw materials which, in turn, are also poorly packaged, in direct contact with the ground and can be lost to the weather. The work takes place in an environment where there is no defined layout for production, waste and inputs, i.e. everything happens in the same place, in a disorganised way (Figures 22 and 23).

Figure 19 - Organisation of inputs.   Figure 20 - Organisation of the RCC.

Source: Own Authorship (Sep/2022)

Transport and disposal are carried out by outsourced rubble companies without any monitoring or tracking of the disposal site (Figure 24).

Figure 21 - RCC Management Flow - Company A.

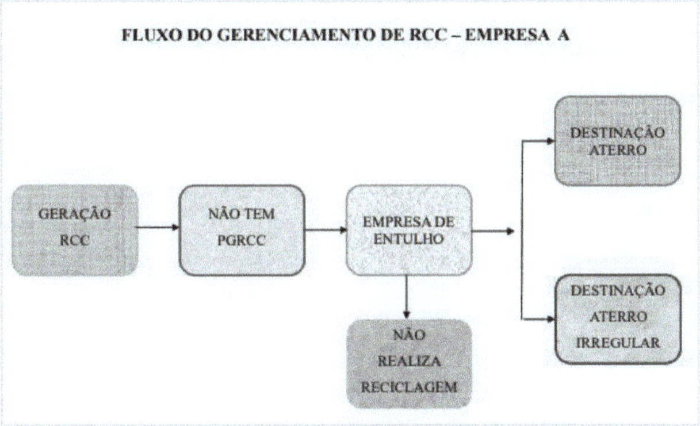

Source: Own authorship (2023)

All waste, regardless of its classification, is stored in 5 metre metal skips[3] . Collection, transport and disposal are carried out by rubble companies, without any kind of treatment, with the aim of recycling the waste. The final disposal takes place in landfill sites or vacant lots, which are favourable for improper disposal. There is no supervision by the administration, except through complaints.

With regard to environmental education on site, the company's employees have no knowledge of the legislation on waste, nor do they receive any kind of training on the importance and need to manage RCC correctly, with a view to minimising the environmental impacts caused by the incorrect disposal of rubble, using the 3R's policy.

Company B - Structural Concrete Block

Object: Construction of a 14-storey residential condominium, leisure area and swimming pool.

Location: Ponta Negra neighbourhood.

A visit to company B's construction site revealed that the building is being built using an unconventional, modular construction system, using structural masonry and concrete blocks. The walls perform a structural function, supporting the entire weight of the structure and distributing it to the foundation, making the use of beams and pillars unnecessary, reducing the use of reinforcement and forms (Figure 25).

Figure 22 - Structural Masonry Construction with Concrete Blocks - Company B.

In addition to supporting the various loads of a building, structural masonry walls have the function of being thermal and acoustic insulators. Among the main advantages of replacing conventional masonry with structural masonry is the reduction in waste and the reduction of waste generation and disposal. Other advantages include:

- Time-saving: The construction time is considerably shorter compared to conventional masonry, since there is no need to make metal frames and structures;

- Costs: Given that certain service steps in conventional masonry are not necessary in this system, there is a reduction in materials and, consequently, in the cost of the work; reducing the use of reinforcement and forms.

- Organisation: It is possible to organise more efficiently, making the environment cleaner and visibly more organised on a construction site, which is essential for the development of the project.

It was observed that the company's employees separate the classified waste into properly identified bays, keeping the construction site clean and organised, thus facilitating proper disposal by the outsourced company. Every day, at morning meetings (DDS), before starting work, employees receive training on the correct handling of waste.

During the planning phase of the project, the company did a lot of research into the correct destination, with a view to meeting the priority objectives of the Laws, Standards and Resolutions on the management of RCC.

Manaus does not yet have a recycling plant or any actions to encourage sustainability in the area of construction waste. Therefore, the solution found was to adopt the modular construction system, using structural block masonry, which at the same time as complying with legislation, also minimises environmental impacts, due to the small volume of waste that the system produces, compared to the conventional system.

Company B monitors the amount of waste transported using a digital platform linked directly to the Ministry of the Environment. The platform tracks the waste generated throughout the production chain. In this way, the process is always in compliance with legislation.

The collection and transport system is carried out by an outsourced company (transporter), which must be registered on the platform used by company B, and must have its licences up to date and be able to carry out these activities. The Waste Transport Manifest (MTR), containing at least the information on quantity, type of waste, volume, company name and destination, must be printed, dated and signed, like an invoice, to accompany the waste to its destination. The destination must also be licensed. Once the companies receiving the solid waste and/or tailings have confirmed receipt of the MTR, the Certificate of Final Destination (CDF) will be available for issue (Figure 26).

Figure 23 - RCC Management Flow - Company B.

Source: Own Authorship (2023)

Company C

Object: Refurbishment of a shed (with mezzanine) to house various shops

Location: Industrial District.

Figure 27 shows the construction of the superstructure. The beams and pillars were made from materials that had been characterised, segregated and sorted for reuse. As this is a refurbishment, company C first classifies what could be reused, recycled and disposed of correctly, demonstrating that the company really does put its PGRCC into practice.

Figure 28, on the other hand, shows the construction site, with the inputs separated from the waste, awaiting the correct destination for both. Despite adopting the conventional construction system, the company applies the forms of waste prioritisation, as established by law, with the aim of minimising waste.

Figure 24 - Superstructure execution.

Figure 25 - Construction site - Company C

Source: Own Authorship (Feb-2023)

The wood is sorted and sent for reuse in other services that need it or for donation to co-operatives. Whenever possible, steel goes through a recycling process to be reused. The rest of the sorted materials are stored in separate bays and, on a weekly basis, are transported by the construction company itself to their final destination in a licensed area owned by the construction company itself, awaiting their reuse in other works by the company or other interested companies. The area, owned by the company, receives waste from its construction sites, as well as from other companies that need to dispose of their rubble.

The company has areas for temporary deposits of Class A waste, which are stored in bays and then recycled using its own machinery (Figure 29). The recycled aggregates are used in the company's own works or those of others. They are used, for example, to produce concrete posts (Figure 30).

Figure 26 - Recycling plant and recycled aggregates.

Figure 270 - Concrete Bollards Produced with Recycled Aggregate.

Photos: Own Authorship (Jan-2023)

The contribution to waste recycling on construction sites relies on the unity and awareness of the entire company team. The construction company benefits from the mobilisation of its entire team, as well as its employees, with actions such as educational campaigns involving talks and staff training. Employees receive guidance on the importance for the environment, as well as for the company, of reducing, reusing and recycling RCC. The concern is much greater about the cost, but in any case there will be a positive impact on the environment.

Company C is an old and pioneering company in Manaus that recycles RCC, not just class A, and several of its construction projects were carried out using recycled construction aggregates. The company also carries out major demolition work in the city, including the collection and environmentally appropriate disposal of this waste (Figure 31).

Figure 28 - RCC Management Flow - Company C.

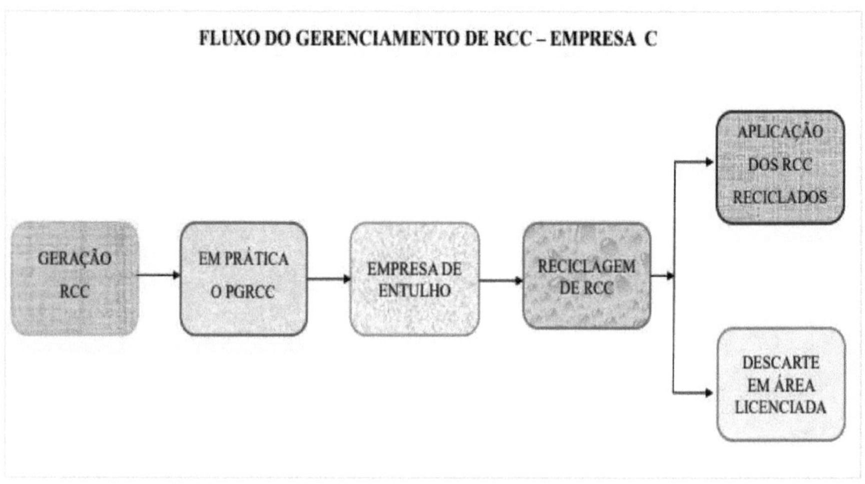

Source: Own Authorship (2023)

Chapter XVI - Weaknesses in the management of construction waste in the city of Manaus

Municipal Construction Waste Management Plan

According to Article 5 of CONAMA Resolution 448 of 18 January 2012, the Municipal Construction Waste Management Plan is the instrument for implementing construction waste management, and must be drawn up by municipalities and the Federal District, in line with the Municipal Integrated Solid Waste Management Plan. The Resolution establishes the obligation for the public authorities to implement the PMGRCC, containing its technical guidelines and procedures for exercising the responsibilities of small and large generators and transporters, as well as areas licensed and made available for the management of RCC.

The absence of a Municipal Solid Waste Management Plan in Manaus hinders the management process by the public administration. In order to change the current scenario, it is necessary to re-evaluate the targets previously set and adapt them to the current reality, with a survey of current figures for generation, collection, transport and final disposal, with a view to complying with the legislation on the subject, as well as urgent, innovative and mitigating actions in favour of the environment.

Licensed areas available for final disposal

The city of Manaus has no registered areas, public or private, suitable for receiving, sorting and temporarily storing small volumes, which would enable the waste to be subsequently sent to processing areas. This favours the emergence of clandestine and illegal landfills in areas suitable for incorrect disposal, at no cost to the generators or transporters. The lack of appropriate sites makes it difficult to reuse and recycle, which makes it impossible to return this waste to the market to be reinserted into the construction chain.

Sustainability in Construction - RCC Recycling

Manaus does not have any public or private recycling plants for later use as raw materials. There are also no financial or tax incentives, through programmes or projects, for new businesses focused on this activity, thus generating jobs and income for the population, as well as benefits for the environment.

The lack of basic infrastructure causes serious environmental, social and economic problems and makes efficient waste management impossible.

Control of the Quantity Generated and the Quantity Disposed of RCC

In the city of Manaus, waste production increases substantially every year and the deficit in waste management is reflected in the city's daily life as a result of the inadequate disposal of solid urban waste, which is visible in various parts of the city, especially in streams and green areas. Manaus does not have access to any control of the volume of waste disposed of, which makes it difficult to find out which areas have been degraded as a result of waste disposal.

> "The imbalance between the quantities discarded and reused makes the management of solid urban waste one of the most serious environmental problems today" (GUARNIERI, 2016).

Control, Monitoring and Inspection of Areas Unsuitable for Disposal

Secretariats and public administration bodies do not have a system for monitoring, inspecting, tracking or mapping the areas used as irregular dumps, which generally consist of legally protected environments with high environmental fragility and precarious environmental control and inspection, favouring occupation. The lack of appropriate places for the correct handling of CCC, the absence of control over disposal and the lack of awareness among the population are the main factors that lead to the creation of irregular waste dumps in the urban centres of capital cities.

Communication between the Secretariats/Bodies of the Administration Involved in the CCW Management Process

The public administration of the city of Manaus does not have any common system or digital platform between the entities involved in the process of managing RCC, such as: secretariats, agencies, transport companies, construction companies and companies that receive rubble, which facilitates the collection of data for decision-making regarding waste management, given that in the current scenario, due to the lack of knowledge of the information, there is no way to manage it.

Supervision of the disposal of RCC generated by construction companies, through Management Plans or Construction Licences issued by Public Administration bodies

There is no control or monitoring of the management of CCW declared by generators in their CCW Management Plans, or of the waste generated by construction companies through the licences issued by public administration bodies. In practice, many generators are unaware of or ignore current legislation.

Monitoring and Inspection of Debris Transport Companies

There are no inspections of rubble disposal companies, except through complaints. The study shows that RCC is most often taken from construction sites and disposed of in inappropriate places. The responsibility for proper disposal, according to Law 12.305/2010, should lie with the generators, but it is transferred to rubble haulage companies. In many situations, the generators do not seek knowledge of the sites used by the outsourced companies for disposal, do not control the transport and correct destination of the RCC, and therefore outsource their responsibilities.

Mapping Green Areas prone to Clandestine Disposal

There is no mapping of the city's green areas, or signposting prohibiting disposal, emphasising the importance of keeping the site free of any unsuitable waste, or the possible imposition of a fine if the offender is identified. This lack of monitoring gives the offender the security that no harm will come to them.

Educational Programmes on CCR and Sustainability

Inadequate waste disposal is common practice in society, proving the lack of educational programmes aimed at sustainability in the construction industry, aimed at training the workforce in how to handle waste correctly and sustainably. The information is usually transmitted in daily safety dialogues (DDS) in a short space of time.

The majority of construction companies do not have qualified professionals to take responsibility for the technical activities in the environmental area, in order to fulfil the main objective of the legislation, in terms of the hierarchy of RCC management.

The RCC collection and disposal companies in the city of Manaus are not registered with CREA-AM and do not have a qualified professional on their staff to monitor the final disposal of RCC and to take responsibility for mitigating the possible negative environmental impacts caused by irregular disposal.

Chapter XVII - Final considerations

It is understood that in terms of environmentally appropriate management of construction waste, the municipality of Manaus is not in compliance with the guidelines established by CONAMA Resolution 307/2002. The municipality approved the Manaus Solid Waste Master Plan (PDRSM) by means of Municipal Law 09 November 2011, published by means of Decree no. 1.349/2011, taking into account Federal Law no. 11.445/2007 and Law no. 12.305/2010, covering technical, environmental, economic and social issues, such as improving the waste collection and treatment infrastructure network, reducing the generation of solid waste, as well as encouraging reuse, recovery and recycling, promoting the economic sustainability of the waste management model, formalising, training, professionalising and fully integrating the informal sector in waste management. The Manaus Solid Waste Master Plan was structured for a 20-year time horizon, i.e. from 2009 to 2029. In the current scenario, the Municipal Solid Waste Plan is being revised, but is not expected to be approved. As a result, Manaus does not have a PMGRCC specifically aimed at construction waste, and it is not possible to demand that the agents involved in the RCC management system, such as generators, transporters and receivers, exercise their responsibilities properly.

With regard to generators, it was found that company A does not have a PGRCC, and its responsibility for waste disposal is transferred to rubble companies. Company B has opted to use a technology that generates less waste and exercises its responsibility for disposal in a correct way, through an accredited company on a digital platform, making it possible to track disposal, company name, quantity, etc. However, this is a requirement of funding bodies. Company C segregates, sorts, correctly disposes of and recycles its waste in its own facilities. What can be seen is that if there are no charges or inspections, companies don't manage their waste properly. In general, the PGRCC submitted to the environmental agencies with a view to obtaining an environmental licence is neither followed nor inspected; it is merely a formality for obtaining the licence. On the other hand, the lack of structure on the part of the municipality means that there are few demands on generators, transporters and receivers. The municipality does not currently have any licensed areas for transhipment and sorting, or for the final disposal of RCC.

In addition, there is no action to promote the reuse and recycling of RCC, or programmes to encourage new entrepreneurs to generate jobs and income.

With regard to the collection and transport of RCC, it was found that Manaus has a sanitary landfill, which receives a minimum amount of RCC through registered companies. However, this does not represent reality, given that there is no way of controlling the quantity or destination of the CDW carried out by the rubble companies. There is no control, monitoring or inspection of the transport companies, which often facilitates disposal in inappropriate places, such as vacant lots, green areas or streams, generating various problems such as negative visual impact, degradation of green areas, accumulation of other waste, proliferation of disease-transmitting vectors, among others.

There is no requirement for qualified professionals to carry out environmental activities in rubble companies or construction companies, with a view to the correct management of RCC.

It was observed that there is no oversight of all the entities involved in the waste management process, as actions are fragmented and there is no information system to enable control and the search for information. In this sense, it is impossible to hold any of those involved accountable, as there is no way of tracking activities. Due to the city's lack of a minimum structure to deal with RCC, it is impossible to supervise and control the actions of the agents involved.

The onus for dealing with issues relating to construction waste cannot be placed solely on the generators, and there is a clear need for the municipality to contribute by defining suitable places to dispose of construction waste, such as ecopoints for small and large volumes, and by introducing environmental education programmes, through partnerships with CREA-AM and Sinduscon-AM, aimed at environmental education on the management and correct disposal of waste, impacts on the environment and sustainability in the construction industry. The creation of a management centre, with its own structure, including projects to inspect construction companies, transport companies and monitor green areas, as well as a hotline to combat illegal disposal, could be important programmes to deal with the entire chain from generation to disposal of RCC in a more efficient way. Policies to encourage recycling on construction sites, through small plants, the establishment of a recycling plant in the municipality, with a view to using recycled aggregates in the construction of affordable housing, for road maintenance, the production of kerbs, in short, there are many applications for recycled aggregates, with a view to complying with legislation in favour of the environment. Sustainability in the construction industry, if used in accordance with the hierarchical order of waste priorities: non-generation, reduction, reuse, recycling and final disposal, brings numerous benefits to the environment, reducing negative impacts by reducing the generation and disposal of waste, reducing the use of natural resources, reducing the cost of environmental remediation, reducing health costs, reducing illnesses, among others.

In addition to reverse logistics applied to construction, it is important to emphasise that efficient management, from project design to construction execution, are ways of reducing waste on site, which contributes significantly to reducing waste generation. The use of new technologies such as the BIM platform, industrialised construction systems and company certifications are also important ways of minimising negative impacts.

Manaus needs to put legislation into practice, adopt national models such as Belo Horizonte, Fortaleza, São José dos Campos and international models that are benchmarks in waste management from a sustainability perspective, such as those used in Holland, Denmark and Germany, because all three powers have in common a strong incentive from the authorities and the construction industry to use reusable/recyclable materials. One of the factors that most influences this high recycling rate is the presence of a tax on the waste generated. There is also a financial incentive from the government to use reusable aggregates, and the high costs attributed to this tax have led CCW generators to prioritise reduction and recycling actions. Public policy actions are therefore urgently needed to structure the CCW management system and thus make the city of Manaus more sustainable, bringing better health and well-being conditions for the population, improved quality of natural resources and better aesthetic and sanitary conditions for the environment.

REFERENCES

ABNT - Brazilian Association of Technical Standards. Classification of Solid Waste, Rio de Janeiro. NBR 10.004, 2004a.

ABRELPE - Brazilian Association of Public Cleaning and Special Waste. Panorama of solid waste in Brazil 2022. Abrelpe, 2022. Available at: Accessed on 17 March 2023.

ALMEIDA, Marcelo Vasconcelos de; SANTOS, Robson de Souza; SILVA, Marcos Meurer da. Analysing Processes for Managing Small Construction Projects: A Case Study in an Architecture Office. Federal University of Itajubá, https://www.redalyc.org/journal/5606/560662203028/html/. Minas Gerais, 2019.

ALVES, RODRIGO COUTO; SILVA, NELITO MARQUES; ANDRADE, MARCOS VINÍCIUS BARROS; MARQUES, EVELY LARANJEIRAS. Municipal Solid Waste Management in Amazonas, Brazil. **Research, Society and Development,** v. 9, n. 12, pp. 1-20, Federal University of Amazonas (UFAM). Manaus, 2020.

BARROS, MURILLO VETRONI. Construction solid waste management plan: an analysis based on CONAMA Resolution 307. **Revista Gestão Industrial,** Ponta Grossa, v. 13, n. 1, p. 139-153, Oct./Dec. 2017.

BATISTA, MARCELO LOPES. Waste Management in Civil Construction: Emphasis on Sustainable Development. **Brazilian Journal of Development,** v. 8, n. 4, pp.23356-23373, Curitiba, April, 2022.

BESEN, Dyanine Weiss; SILVA, Rhaiza Lima Maia da. Innovative Technologies and Sustainability in Civil Construction: A Case Study in Santa Catarina, SC. Course Conclusion Paper presented to the Civil Engineering Course at the University of Southern Santa Catarina, Palhoça, 2017.

BRAGA, Larissa Passos; VEIGA, Thais da Costa. Analysing Construction and Demolition Waste Management on construction sites in the Federal District. Final

Project Monograph - University of Brasília. Faculty of Technology. Federal District, 2017.

BRAZIL. Law no. 12.305 of 2nd August 2010. **National Solid Waste Policy. Brasília, DF, 2010.**

BRUM, Fábio Martins. **Implementing a Construction Waste Management Programme on a Public Construction Site: The case of UFJF.** 2013. 107f. Dissertation presented to the Postgraduate Programme in the Built Environment at the Federal University of Juiz de Fora, Minas Gerais, 2013.

CASIRINI, Lucas. Optimisation of the Solid Waste Management Plan: Case study: VITRIUM - Intelligent Medical Centre. Conclusion of Lato Sensu Postgraduate Course in Environmental Analysis and Sustainable Development. University Centre of Brasília (UniCEUB/ICPD). Brasília, 2015.

CAMENAR, Mariana Thays; SCHEID, Melquior Forgiarini. Analysing the Construction Waste Management System: Case Study in the Municipality of Pato Branco - PR. Graduation work presented to the Civil Engineering course at the Federal Technological University of Paraná - Pato Branco Campus, 2016.

CESÁRIO, Jonas Magno dos Santos. Et al. **Scientific methodology: Main types of research and their characteristics.** Multidisciplinary Scientific Journal Núcleo do Conhecimento. Year 05, Ed. 11, Vol. 05, pp. 23-33. November 2020, Access link: https://www.nucleodoconhecimento.com.br/educacao/tipos-de-pesquisas.

CIDADE, FERNANDA CABRAL; OLIVEIRA, JOSÉ ALDEMIR DE. From Collection to Commercialisation: The Reverse Production Chain of Post-Consumer Goods in an Amazonian City. **Geo UERJ,** Rio de Janeiro, n. 31, pp. 474-503, 2017.

CONAMA - National Environment Council. Resolution no. 307, of 5 July 2002. Brasília. Diário Oficial da União, 30 August 2002, section I, p. 17.241.

CONKE, LEONARDO SILVEIRA; NASCIMENTO, ELIMAR PINHEIRO. Selective Collection in Brazilian Surveys: A Methodological Assessment. ***Brazilian Journal of Urban* Management**, Jan/Apr, 10(1), pp. 199-212, Brasília, 2018.

COSTA, Bianca da Silva Lima Miconi. A Study on Sustainability. Monograph presented to the Specialisation Course in Production and Management of the Built Environment at the School of Engineering of the Federal University of Minas Gerais, Belo Horizonte, 2019.

CRUZ, ZOZIMERE DO CARMO DA SILVA. Solid Construction Waste in Manaus. **Specialise ON-LINE IPOG Magazine.** Year 9, ed. 16, v. 1, Goiânia, December, 2018.

DINIZ, BRUNA LETÍCIA SOARES. The Importance of Correct Disposal and Recycling of CSR in the Municipality of Manaus. Proceedings **of the VII SINGEP** (International Symposium on Project Management, Innovation and Sustainability - São Paulo - SP - Brazil - 22 and 23/10/2018.

EIGENHEER, EMÍLIO MACIEL. The History of Rubbish. Urban Cleaning through the Ages. Rio de Janeiro, ELS2, 2009.

FONSECA, MARIA JULIA M.; CASTRO, MARCUS AVEZUM A. DE; MAINTINGUER SANDRA I. Paper presented to the Postgraduate Programme in Regional and Territorial Development - Federal University of Araraquara (UNIARA). Application of Reverse Logistics in Civil Construction as a Sustainable Environmental Mechanism in Public Policies. Araraquara - SP, 2018.

GRUBLER, T. H. Comparative study between the Light Steel Frame, conventional masonry and structural masonry construction methods. 2021. Final Course Work (Bachelor's Degree in Civil Engineering) - Universidade Regional do Noroeste do Estado do Rio Grande do Sul, RS, 2021.

JACINTO, ANA CAROLINA. Public Policy on Solid Waste: An analysis of Law 12.305/2010, which establishes the National Policy on Solid Waste - PNRS, through the services performed by the Municipal Department of Public Cleaning - SEMULSP in the Municipality of Manaus. **Revista de Administração de Roraima, UFRR-Boa Vista,** v. 6, n. 2, pp. 510-520, Boa Vista, 2016.**

JUNIOR, AIRTON BRAGA TEIXEIRA; PRADO, DANIELLE APARECIDA DO; JUNIOR, MILTON GONÇALVES DA; PARADA, JOAQUIM ORLANDO; FELICIANO AURÉLIO CAETANO; PEREIRA, RAFAEL GONÇALVES FAGUNDES; DIAS, JÉSSICA. Study of the Reuse and Recycling of Solid Construction Waste. **Uniaraguaia Magazine.** V. 15, n. 1, pp. 1-23, Goiânia, 2020.

LIMA, Bruna Pietroski de; Management of Construction Waste in the Municipality of Maringá-PR. Article presented to the Undergraduate Course in Civil Engineering at UNICESUMAR - Centro Universitário de Maringá, 2019.

MORAND, Fernanda Guerra. Study of the Main Applications of Construction Waste as Building Materials. Graduation project presented to the Civil Engineering course at the Polytechnic School, Federal University of Rio de Janeiro, Rio de Janeiro, 2016.

MOTA, ANTONIO RONEY DA; SILVA, NELITON MARQUES DA. Historical Scenario and Considerations on Solid Waste. **Revista Desarrollo Local Sostenible**. V.7, n. 20, pp. 1-18, Amazonas, 2014.

MOTA, Jeane. **The Disposal of Solid Construction Waste in Manaus: From the Construction Site to the Final Destination**. 2014. 72f. Master's dissertation from the Institute of Technology Professional Master's Degree in Construction Processes and Urban Sanitation, Belém-PA, 2014.

NASCIMENTO, Yan Carlos. Diagnosis of Municipal Construction Waste Management in the Municipality of Toledo -PR According to CONAMA

Resolution 307/2002. Conclusion paper for the Civil Engineering course at the Federal Technological University of Paraná. Federal Technological University of Paraná (UTPR), Toledo (PR), 2018.

NAZARI, MATEUS TORRES; GONÇALVES, CAROLINA DA SILVA; DALL'AGNOL, ANA LUIZA BERTANI; SILVA, PAMELA LAIS CABRAL; REGINATTO, CLEOMAR. Evolution of Brazilian Environmental Legislation on Solid Waste. **2nd South American Congress on Solid Waste and Sustainability (ConReSol)**, Foz do Iguaçu - PR, 28-30 May 2019.

NOBRE, Lucas Casarini. Optimising Solid Waste Management: VITRIUM-Intelligent Medical Centre Case Study. Conclusion paper for the *Lato Sensu* Postgraduate course in Environmental Analysis and Sustainable Development, University Centre of Brasília (UniCEUB/ICPD), Brasília, 2015.

NOBRE, Paulo Lobato. Analysing some of BIM's contributions to Construction Planning and Control. Monograph presented to the Specialisation Course in Management and Technology in Civil Construction at the UFMG School of Engineering, Belo Horizonte, 2021.

NOGUEIRA, CRISTIANY DA SILVA. Construction Waste Management. **Multidisciplinary Scientific Journal Núcleo do Conhecimento**. Year 05, ed. 11. v. 10, pp. 67-84. Manaus, November 2020.

NUNES, Gabriel de Freitas; SOUZA, Patrick Fhilippi de. Potential for Reducing Waste in Construction by Replacing Conventional Building Systems with Industrialised Systems. Course Conclusion Paper presented to the Civil Engineering Course at the Universidade do Sul de Santa Catarina - UNISUL, Tubarão, 2017.

OLIVEIRA, MARIA DO P.S. LÂMEGO; OLIVEIRA, EVAILTON ARANTES DE; CAMPOS, FILHO, LUCIANO MOREIRA DE SOUZA; TEIXEIRA, ANA MARGARIDA M. FONSECA. Environmental Implications of the Final

Destination of Construction Waste in the Municipality of Manaus (Amazonas-Brazil). **Scientific Technical Congress of Engineering and Agronomy (CONTECC),** Maceió, AL, 21-24 August 2018.

OLIVEIRA, MARIA DO P.S. LÂMEGO; OLIVEIRA, EVAILTON ARANTES DE; CAMPOS, ARLENE M. LÂMEGO DA S.; FONSECA, ANA MARGARIDA. Characterisation of Waste on Construction Sites in the City of Manaus. **Scientific Technical Congress of Engineering and Agronomy (CONTECC),** Palmas, TO, 17-19 September 2019.

OLIVEIRA, Maria do Perpétuo Socorro Oliveira. **Development of Strategies for the Management of Construction and Demolition Waste in the Municipality of Manaus (Amazonas-Brazil) based on the Circular Economy concept**. 2021. 98f. Thesis (Doctorate in Ecology and Environmental Health) - Fernando Pessoa University, Porto, 2021.

OLIVEIRA, Rômulo de Lima de Management of Construction Waste - Intelligent Solutions. Undergraduate project presented to the Civil Engineering programme at the Federal University of Santa Maria (UFSM, RS), 2018.

OLIVEIRA, F. de A.; MAUÉS, L. M. F.; ROSA, C. C. N.; SANTOS, D. de G.; SEIXAS, R. de M. Predicting waste generation in construction using BIM modelling. **Ambiente Construído**, Porto Alegre, v. 20, n. 4, p. 157-176, Oct./Dec. 2020.

STATE SOLID WASTE PLAN FOR AMAZON - PERS-AM; Stage VI - Task II; Product III, Sep/2017.

MANAUS SOLID WASTE MASTERPLAN (PDRSM, 2010).

MANAUS MUNICIPAL INTEGRATED SOLID WASTE MANAGEMENT PLAN (PMGIRS), Public Consultation version, 2015.

NATIONAL SOLID WASTE PLAN, 2022.

PORTO, Júnia de Oliveira. **Development of an Integrated Urban Solid Waste Management Evaluation System: Application to the southern region of Ride/DF and its surroundings.** 2017. 250f. Master's dissertationUniversity of Brasília. Faculty of Technology, Federal District, 2017.

PEREIRA, Emiliano Santos. **Civil Construction at UFAM: A Proposal for Waste Reduction.** 84f. Dissertation presented to the Postgraduate Programme in Production Engineering at the Federal University of Amazonas, Manaus, 2017.

PEREIRA, Flávia Angélica dos Santos Martiniano. Reusing Construction and Demolition Waste: A Sustainable Construction Model. Graduation project presented to the Civil Engineering course at the João Pessoa University Centre, 2018.

PEREIRA, UILANE DE AMORIM. Urban Solid Waste as a Conditioner of Diseases in the City of Manaus (UFAM). **Revista Geonorte,** v.9, n. 31, pp. 32-53, Manaus, 2018.

PINHEIRO, Mayrla Abreu. Analysing the Use of Unconventional Tools in Civil Construction. Graduation project presented to the Civil Engineering course at the Federal Institute of Education, Science and Technology of Paraíba, Campus Cajazeiras, PB, 2021.

National Solid Waste Plan. https://www.gov.br/mma/pt-br/noticias/governo-federal-acaba-com-a-espera-de-mais-de-10-anos-e-publica-decreto-do-plano-nacional-de-residuos-solidos#:~:text=After%20more%20than%202010,valid%20in%20all%20territ%C3%B3rio%20national.

POZZETTI, VALMIR CÉSAR; CALDAS, JEFERSON NEPUMUCENO. Solid waste disposal at the heart of sustainability. **Revista de Direito Econômico e Socioambiental**, v. 10, n. 1, p. 183-205, Curitiba, Jan./Apr. 2019.

RODRIGUES, Fernanda Costa, WANDEREI, JOSÉ PEDRO DE CARVALHO. Reverse Logistics and Corporate Sustainability: Case Study at ZIMERPLAS. Graduation project presented to the Production Engineering course at Instituto Vale do Cricaré, São Mateus, ES, 2019.

SÁ, MARCOS VINÍCIUS OLIVEIRA DE; MALHEIROS, ALEXANDRE JOSÉ DE AANDRADE; SANTANA, CLAUDEMIR GOMES DE. The Importance of CONAMA Resolution 307 for the Management of Solid Construction Waste. **Journal of the Centre for Sustainable Development Studies (CEDS)**, n. 9, August-Dec, 2018.

SERINOLLI, GUSTAVO PERPÉTUO. Construction Waste: A New Look at Recycling. **Scientific Journal Academic Week Fortaleza-CE**, v. 9, issue 205, pp. 01-17, 2021.

SCHWENGBER, Estela Regina. Construction Waste. Course Conclusion Paper presented at the Specialisation Course in National and International Environmental Law, Federal University of Rio Grande do Sul (UFRGS), 2015.

SILVA, ÁDRIA SOUZA DA; SANTOS, ROBERTA MONIQUE DA SILVA; VIANA, ÁLEFE LOPES; CARNEIRO, CAMILLA JACQUELINE MEDEIROS; SILVA, PRISCILA THAYANE DE CARVALHO; SANTOS, KAYRA JORDANA SÁ DOS; LACERDA, FRANCISCO ANTÔNIO SIEBRA; FREITAS, CINTHIA RÉGIA DOS SANTOS. Solid waste management in civil construction: A case study of two companies in the city of Manaus-AM. **InterfacEHS-Saúde, Meio Ambiente e Sustentabilidade,** Centro Universitário Senac. v. 12, n. 1, pp. 56-67, São Paulo, 2017.

SILVA, OTÁVIO HENRIQUE DA; UMADA, MURILO KEITH; POLASTRI, PAULA; NETO, GENEROSO DE ANGELIS; ANGELIS, BRUNO LUIZ DOMINGOS DE; MIOTTO, JOSÉ LUIZ. Steps in the Management of

Construction Waste. **Electronic Journal on Environmental Management, Education and Technology**, UFSM. V. 19, pp. 39-48, Santa Maria, 2015.

SOUZA, Kelen Gomes de. **Solid Waste in the city of Manaus.** 2014. 64f. Master's dissertation from the Institute of Technology Professional Master's Degree in Construction Processes and Urban Sanitation, Belém-PA, 2014.

TAVARES, QUEZIA ELAINE DA SILVA; SANCHES, ANTONIO ESTANISLAU; BANDEIRA, SANDY REBELO; MARQUES, DANIEL DA SILVA; SANTOS, GLEICINÁRIA OLIVEIRA DOS. Identification of irregular disposal sites for construction and demolition waste in the industrial district II neighbourhood in the municipality of Manaus-AM. **Braz. J. of Develop.**, Curitiba, v. 6, n. 2, p. 6014-6024, feb. 2020.

UN Environment Programme: https://www.unep.org/pt-br/noticias-e-reportagens/reportagem/o-que-voce-precisa-saber-sobre-estocolmo50. Accessed on 12 September 2022.

SEMULSP, 2022. Available at: ANNUAL REPORT 2021.docx (manaus.am.gov.br). Accessed on: 17/08/2022.

yes
I want morebooks!

Buy your books fast and straightforward online - at one of world's fastest growing online book stores! Environmentally sound due to Print-on-Demand technologies.

Buy your books online at
www.morebooks.shop

Kaufen Sie Ihre Bücher schnell und unkompliziert online – auf einer der am schnellsten wachsenden Buchhandelsplattformen weltweit! Dank Print-On-Demand umwelt- und ressourcenschonend produzi ert.

Bücher schneller online kaufen
www.morebooks.shop

info@omniscriptum.com
www.omniscriptum.com

Printed by Books on Demand GmbH, Norderstedt / Germany